AF179547

Liability of the Negligent Surveyor

Written by Ben Maltz LLB(Hons), Barrister
Edited by Victor H Vegoda FRICS IRRV FBEng PEng

Number 10 in the "Need to Know" Series

Copyright Lark Productions Limited and Estates Gazette

Jointly Published by

Estates Gazette
151 Wardour Street
London W1

and

Lark Productions Limited
PO Box 2933, London N20 9LR

Published February 2001

ISBN Number 1 898383 65 0

Other titles in the "Need to Know" Series:-

Practical Ways to Recover your Debts;
A Practical Guide to Residential Surveys;
Schedules of Dilapidations;
Heave, Demolition & Reconstruction;
Draft Your Own Terms of Engagement;
Condam – Draft Your Own Safety Plan;
Landlord & Tenant Covenants Act, 1995;
An Introduction to the Party Wall etc. Act, 1996;
Contracts (Rights of Third Parties) Act 1999.

Printed by Bell & Bain Ltd., Glasgow

Contents

<u>WARNING!</u>

This book is written on the basis of the law in England and Wales. Property law in Scotland differs from that of the rest of the United Kingdom. This book is intended as a quick reference guide to the law relating to surveyors' negligence. The reader should refer to authoritative texts on the subject such as Jackson & Powell on Professional Negligence for fuller discussions on the issues raised in this work.

This book sets out the position of the law in the field of surveyors' negligence as it stands at 30th June 2000.

Chapter One

The Tort of Negligence and its Development in English Law

A tort in English law may be regarded as a 'civil wrong' and negligence has long been recognised, by lawyers and academics alike, as the one predominant tort which accounts for the vast majority of tortious litigation to pass through the English civil courts. The tort of negligence may be defined as 'the breach of a legal duty to take care which results in damage, undesired by the defendant, to the plaintiff.'[1]

In order for a claim in negligence to succeed the claimant must prove three exact elements.[2] The first element is to establish that the defendant owed the claimant a *'duty of care'*. Secondly, it must be shown that the defendant *'breached'* that duty of care and finally the claimant must prove that he has suffered *'damage'* as a result of the defendant's breach of his duty of care.

[1] Winfield & Jolowicz on Tort, 14th ed. p. 78
[2] Lord Wright in Lochgelly Iron & Coal Co v. M'Mullan [1934] AC 1

This book will consider the application of these three elements to the specialist area of surveyors' negligence. However, in order to understand the fundamental concepts of the subject, it is first necessary to analyse the development of each of the three elements in terms of basic negligence claims and the way in which the courts have interpreted them.

Establishing a Duty of care

It can be said that the birth of the modern day negligence action arrived with the advent of the landmark case of **Donoghue v. Stevenson [1932]**[3] in which Lord Atkin sought to formulate a general tortious principle in an attempt to alleviate the constraints imposed upon potential claimants by the doctrine of privity of contract. Lord Atkin stated that:

> *"The liability for negligence, whether you style it such or treat it as in other systems as a species of 'culpa', is no doubt based upon a general public sentiment of moral wrongdoing for which the offender must pay.....The rule that you are to love your*

[3] [1932] AC 562

neighbour becomes, in law, you must not injure your neighbour;.....You must take reasonable care to avoid acts or omissions which you can reasonably foresee would be likely to injure your neighbour. Who then, in law, is my neighbour? The answer seems to be - persons who are so closely and directly affected by my act that I ought reasonably to have them in contemplation as being so affected when I am directing my mind to the acts or omissions which are called in question."

The neighbour principle

Lord Atkin's concept, which later became known as the 'neighbour principle', initially faced some considerable criticism. Many judges felt that the **Donoghue** case was confined to its facts and that it merely established the rule that manufacturers, as the defendant was in that case, owed a duty not to cause physical damage to ultimate consumers of their wares. However, Lord Macmillan had noted in **Donoghue** that "the categories of negligence are never closed" and indeed, by 1970 Lord Reid in **Home Office v. Dorset Yacht Co. Ltd [1970]**[4] suggested that the neighbour principle applied in all cases where no valid reason for its exclusion could be given.

The wider application of the neighbour principle was also approved in another House of Lords' case, **Anns v. Merton London Borough Council [1978]**[5]**,** in which Lord Wilberforce set about creating a two stage test in order to determine when a duty of care would arise in any particular situation. The first stage, which Lord Wilberforce referred to as:-

the 'relationship of proximity',

incorporated Lord Atkin's neighbour principle.

The second stage, which he suggested should be considered if:-

a "sufficient" relationship of proximity is established,

involved the court looking towards any considerations

---

[4] [1970] AC 1004
[5] [1978] AC 728

which negated or reduced the scope of the duty.

However, in a series of House of Lords cases which followed **Anns**, including one professional negligence case, Lord Wilberforce's two stage approach or rather the wide application of it was treated with quite considerable reservation[6]. This trend in moving away from the **Anns** approach culminated in the Privy Council case of **Yuen Kun-yeu v. Attorney-General of Hong Kong [1987]**[7] in which Lord Keith suggested that the two-stage test formulated by Lord Wilberforce had been 'elevated to a degree of importance greater than its merits, and greater perhaps than its author intended.'

Lord Keith was of the opinion that the second stage would rarely be applicable[8] but more importantly he was clear to point out that the first stage carried with it the risk of

---

[6] see Lord Keith in Peabody Donation Fund v. Sir Lindsay Parkinson & Co [1984] 3 All ER 529, Lord Brandon in Leigh & Sullivan Ltd v. Aliakmon Shipping Co Ltd [1986] 2 All ER 145 and Lord Bridge in Curran v. Northern Ireland Co-ownership Housing Association [1987] 2 WLR 1043

[7] [1987] 2 All ER 705

[8] see Hill v. Chief Constable for West Yorkshire [1987] 2 WLR 1126 and
Rondel v. Worsley [1969] 1 AC 191 per Lord Pearce

misinterpretation. Lord Wilberforce's approach could be construed as:-

a measure of sufficiency of proximity

by means of a test of reasonable contemplation or alternatively as a test more in line with Lord Atkin's thinking which incorporates the concept of a necessary relationship between plaintiff and defendant.

In effect, what has transpired is a reversion to 'proximity' as the single test for establishing when a duty of care arises, although it is worthy to recognise that policy considerations are an ever present factor in judicial proceedings.

The decision in **Murphy v. Brentwood District Council [1991]**[9] effectively put the issue to rest by overruling **Anns**. This case is of particular relevance to the area of surveyors' negligence because it involved a claim of negligence against the defendant's surveying department who had approved what were found to be faulty raft foundations in a building which the plaintiff had acquired.

---

[9] [1991] 1 AC 398; [1990] 2 All ER 269

Since the fall of the **Anns** approach the courts have recognised that less emphasis should be placed on policy considerations when considering the duty of care. In the case of **Caparo Industries plc v. Dickman [1990]**[10] the House of Lords formulated what may be described as the most modern approach to duty of care. Lord Oliver suggested that initially the question should be asked 'Is the situation before the court covered by an existing House of Lords case which states that a common law duty of care is owed here?'

Lord Oliver continued by explaining that if the answer to this question was 'yes' then a *prima facie* duty of care is owed. However, if the answer was 'no' Lord Oliver ruled that the following four tests should be applied to the case in question:-

- *was the damage suffered by the plaintiff reasonably foreseeable*

[10] [1990] 2 AC 605; [1990] 1 All ER 568

- *is there sufficient proximity between the plaintiff and defendant that ought to have been in the contemplation of the defendant*

- *is it just and reasonable to impose a duty of care on the defendant*

- *is the imposition of the duty of care a justifiably incremental step*

If all four tests are satisfied then, accordingly, the defendant will owe a duty of care. However, it is interesting to note that this 'modern' approach still encompasses substantial scope for policy considerations as was pointed out by Lord Keith of Kinkel in the case of **Alcock v. Chief Constable of South Yorkshire [1991]**[11]

Breach of duty and Standard of care

It is clear that the ordinary person is capable of owing a

[11] [1992] 1 AC 310; [1991] 4 All ER 907

duty of care to a third party who has been affected by his or her negligent action but what can be said of the responsibility or duty which a professional person incurs. A professional person can owe a duty of care to a third party, indeed it may be argued that there is a greater implication of a duty of care when dealing with professional defendants.

Two stage process to determine whether there has been a breach of duty

It can be said that in many negligence claims, professional or otherwise, the predominant difficulty with which the claimant is faced is not having to prove that a duty of care exists, but establishing that the defendant has breached that duty. It is well established law that the courts apply a two stage process to determine whether there has been a breach of duty:

> *first, what was the standard of care required of that particular defendant; and*
>
> *secondly, on the facts in question did the defendant fall below that required standard.*

It is the first of these questions which separates general negligence actions from professional negligence claims. In ordinary cases of negligence, the test used is that of the reasonable man or to use a modern analogy 'the man on the Clapham Omnibus'. Lord Macmillan in the case of **Glasgow Corporation v. Muir [1943]**[12] sought to define the reasonable man as a man who "is presumed to be free both from over-apprehension and from over-confidence". This concept of reasonable foresight is an objective test and thus the courts will consider whether the ordinary man would have acted in the same way.

The landmark case of **Bolam v. Friern Hospital Management Committee [1957]**[13] was the first case to lay down the test for determining the standard of care in instances of professional negligence. The test remains objective but the relevant question becomes :-

> *'would the ordinary man, who exercises and professes to have*

[12] [1943] 2 All ER 44
[13] [1957] 2 All ER 118

that special skill, have acted in the same way as the defendant did?'

Bolam was a case concerning medical negligence and indeed the medical profession has provided much of the case law in the progression of modern professional negligence.

Standard of care objective not subjective

Without wanting to delve too deeply into the case law surrounding medical negligence, one case in particular discussed further the issue of standard of care for professional defendants. In the case of **Wilsher v. Essex Area Health Authority [1987]**[14] a junior doctor attempted to argue that his own inexperience lowered the standard of care required of him.

However, the Court of Appeal reiterated the fact that the test for assessing standard of care was objective and not, as counsel for the defendants had purported, subjective. The case was subsequently appealed on a causation point but the House of Lords confirmed that the Court of Appeal had been correct in

[14] [1987] 2 WLR 425

their application of the standard of care test.

Not negligent to follow recognised course of action

Medical negligence case law has also established a number of principles in relation to professional practices. In **Ashcroft v. Mersey Regional Health Authority [1983]**[15] it was held by the court that provided a doctor follows a course of action or treatment which is recognised by an established body of the medical profession, he shall not be liable in negligence.

Not negligent to choose one recognised course in preference of another

Furthermore, in the case of **Maynard v. West Midlands Health Authority [1984]**[16] the principle was established that where a doctor or other medical practitioner chooses to follow one recognised treatment in preference of another recognised treatment, once again he will not be held negligent on the basis of this fact alone.

---

[15] [1983] 2 All ER 245
[16] [1984] 1 WLR 634

Causation and Remoteness of damage

The third requirement for a successful negligence claim is the burden of proving that the claimant has suffered damage. This is an area which is primarily concerned with the concepts of causation and remoteness of damage. Causation must be proved so as to establish a causal link between the defendant's breach of duty and the damage suffered by the claimant. Furthermore, the damage complained of by the claimant must be judged by the court as damage which is not too remote. These are two separate issues, both of which have implications for professional negligence cases, and thus they are dealt with independently below.

The 'but for' test
novus actus interveniens

The standard test for establishing causation is known as the 'but for' test as was applied in the case of **Barnett v. Chelsea & Kensington Hospital [1968]**[17]. The court should consider whether the claimant would not have sustained the damage 'but for' the defendant's actions or omissions. In addition to this test, the presence of any external

[17] [1968] 1 All ER 1068

causative factors must be ruled out. Such intervening factors, otherwise known as *novus actus interveniens*, can be said in certain circumstances to 'break' the chain of causation, thus negating the defendant's liability[18].

Some other case law in the area of causation has dealt with scenarios in which there have been more than one possible cause for the damage suffered. It is unclear in which exact circumstances causation will be proved but two particular cases assist in answering this question.

In the case of **McGhee v. National Coal Board [1972]**[19] the presence of two possible causes of the plaintiff's illness was not enough to relieve the defendant of tortious liability. However, more recently in **Wilsher v. Essex Area Health Authority [1987]**[20] it was discovered that the defendant was responsible for only one of six possible causes of the plaintiff's blindness. The House of Lords held that causation could not be proved on these

[18] see The Oropesa [1943] 1 All ER 211 and Knightley v. Johns [1982] 1 WLR 350
[19] [1972] 3 All ER 1008
[20] [1987] 2 WLR 425

facts and thus the Health Authority evaded liability.

Remoteness of damage

Even where a causal link can be established the question of whether the damage is too remote arises. Remoteness of damage is a question of law rather than fact and it can be regarded as a line drawn by the courts at which point the defendant's liability ceases. The traditional test for remoteness of damage was laid down in **Re Polemis and Furness, Withy & Co [1921].**[21] in which Bankes LJ stated 'the anticipations of the person whose negligent act has produced the damage are irrelevant.'

Reasonable foreseeability

However, what has long since been recognised as the correct approach to remoteness of damage is the test in **Overseas Tankship (UK) Ltd v. Morts Dock and Engineering Ltd [1961]**[22]otherwise known as **The Wagon Mound No 1**. This case established that the modern test is one based on reasonable foreseeability and thus the defendant can only be held liable for damage which could

---

[21] [1921] All ER Rep 40
[22] [1961] 1 All ER 404

have been foreseen by that elusive character, the reasonable man. Since **The Wagon Mound No 1**, remoteness of damage has been qualified somewhat further.

The case of **Hughes v. Lord Advocate [1963]**[23] showed that where the 'type' of damage sustained, rather than the extent of the damage sustained, was foreseeable then this was not too remote.

In **Hughes** the House of Lords held that an explosive combustion was a more serious 'type' of fire and thus was forseeable. In contrast, the facts in the case of **Doughty v. Turner Manufacturing Co Ltd [1964]**[24] involved a chemical reaction that occurred when an asbestos cover was inadvertently knocked into a vat of chemicals. It was held that the damage to a worker was too remote as the chemical reaction was not merely a 'type' of splash but generically, a different kind of damage altogether.

[23] [1963] 1 All ER 705
[24] [1964] 1 All ER 98

Chapter Two

An Explanation of Valuation and Structural Surveys

This chapter gives an analytical breakdown of the type of services that a practising surveyor will offer to a house purchaser or mortgage lender. It is essential that the accepted standards of the profession are understood and therefore the various guidelines and standard form issued by the Royal Institution of Chartered Surveyors (RICS) will be examined later in this chapter.

The need for surveys

The majority of people in the United Kingdom will, at some stage in their lives, purchase some form of property, be it a house, flat, maisonette or bungalow. Similarly the majority of people will need to secure a loan, typically a mortgage, in order to purchase that property. Whether a purchaser requires a mortgage or not, one question remains constant in the mind of any potential purchaser:-

'Is the house I am interested in buying structurally sound?

Of course, many purchasers will be interested in knowing the condition of the property to a much greater degree of detail but primarily, every purchaser, whether private or commercial, wants to know that they are not purchasing a financial liability. Structural defects in properties can be very expensive to cure and in cases of severe structural defects repair costs for treatments such as 'underpinning'[25] have been known to exceed the value of the property itself.

The principle of *'caveat emptor'*

It is important to realise that the underlying need for surveys is derived from a contractual principle at common law; the maxim '*caveat emptor*', which is translated as 'let the buyer beware'. This doctrine, which is unique to English law, has the effect of placing a burden on the purchaser of goods to inspect them for any faults prior to purchase. Therefore 'in the absence of fraud and providing the goods were open to inspection, the buyer cannot complain of defects in the goods which he has purchased.'[26]

[25] Underpining is a technique used to reinforce the damaged or defective foundations of a structure
[26] Cheshire Fifoot & Furmston's Law of Contract, 12th ed. p.136

Although there has been subsequent legislation such as **The Sale of Goods Act 1979**[27] and **The Consumer Protection Act 1987**[28], to protect purchasers from the effect of *caveat emptor*, this legislation, does not apply to contracts for the sale of real property. That is, land and any property appurtenant to that land. Therefore, it is not the duty of the vendor of a property to disclose any defects but instead, an obligation of the purchaser. It should be noted that vendors have been held to owe a duty of care to purchasers when they are required to disclose certain information by way of solicitors' pre-contract enquiries. However, this is not an issue directly relevant to the field of surveyors' negligence.

The vast majority of house purchasers require a mortgage. Mortgages can make up as much as 95 per cent of the purchase price of a property and not surprisingly, the lender is keen to obtain security for the loan. A growing number of cases to reach the courts in recent times are brought by the mortgage

---

[27] Sections 14(2) & 14(3) 'Satisfactory quality' and 'Fitness for purpose'

[28] Section 2 'Liability for defective products'

lenders and many of these cases are dealt with later in this book. Security for a mortgage is nearly always by way of a legal charge over the property being purchased, and thus in the event of failure to meet the loan repayments, the lender is entitled to repossess the property and recover what he can from the proceeds of sale.

Every mortgage lender will have a valuation survey carried out before approving a loan. This is to establish that in the event of a forced sale, the property on which the loan is to be secured, will be sufficiently valuable to discharge the debt. It has become common practice for a copy of such a report to be handed to the purchaser. This is why, as demonstrated in **Yianni v. Edwin Evans [1982]**[29], only about 15 per cent of home buyers instruct an independent structural surveyor. It is here that the distinction between a valuation and a structural survey must be made.

The valuation survey

The primary objective of a valuation survey is satisfy the <u>lender</u> (the mortgagee) of the value and condition of the

[29] [1982] 1 QB 438

property and to establish its suitability to be security for the loan. A valuation survey does not go into anywhere near as much detail as a structural survey and hence, in reality, it should not be regarded as a substitute for a structural survey. The lower level of detail expected of a valuation survey by a mortgage lender is reflected in the surveyor's fee. Valuation surveys can be carried out for as little as £50 but generally surveyors charge in the region of £150 with some firms asking 0.1% of the purchase price.

Before explaining the aspects of a typical inspection, the term 'valuation' needs to be qualified because the value may vary depending on the format of the mortgage lender's standard valuation form. The RICS publishes a booklet entitled 'Comments and Guidance on the Standard Conditions of Engagement & Guidance on Reporting'[30]. The 'open market value' is defined in this guide as 'the best price at which the sale of an interest in property might reasonably be expected to have been completed unconditionally for cash consideration on the date of valuation.'

---

[30] Appendix B for a copy of the guide

Some valuations will require a reinstatement value to be assessed, especially when the value is to be used for insurance purposes. Although the reinstatement value is not defined in the RICS guidance booklet, it can be said to be the cost of having to reinstate or replace the property. The booklet states that a surveyor, when considering the reinstatement value of a property, should have regard to the ABI (Association of British Insurers) & BCIS (Building Cost Information Service) publication entitled 'Guide to House Rebuilding Costs'.

The inspection which is carried out during a valuation survey is aimed at assessing the overall condition of the property and determining that there are no defects which would have a significant effect on the value. The main areas covered in a typical mortgage lender's valuation report are outlined below: -

- *General information including the length of the lease (if applicable), a schedule of accommodation (giving brief details of the number and type of rooms in the property) and the construction type of the roof and main walls.*

- *Information on which services are connected or available (i.e. water, drainage, electricity and gas) and whether the adjoining road or roads are privately owned or maintained by the local highway authority.*

- *A statement is usually required on the general state of repair of the property, with particular emphasis on any essential repairs such as the presence of subsidence, heave, landslip or flooding.*

- *An assessment of the property's open market value and reinstatement value, and a comment on whether the property is suitable security for the purposes of the mortgage applied for.*

- *Some valuation reports require a site plan to be drawn in order to give the lender an idea of the layout of the property and its plot.*

The structural survey

A structural survey goes into much greater detail than a valuation and is thus often commissioned by prudent purchasers who realise the inadequacy of the mortgage lender's valuation survey. Generally, structural surveys will cover the same ground as a valuation survey but involve a more detailed 'room by room' inspection. Structural survey reports can be quite lengthy and thus fees are rarely less than £300 and will usually be calculated according to the size of the property.

Structural surveys do not generally include an assessment of the property's open market value or value for insurance purposes. However, a structural survey report will almost certainly have within its pages a schedule of repairs (if appropriate) and the estimated costs of rectifying each of the defects. The main areas are as follows:

- *A detailed description of the property including a schedule of accommodation, descriptions of any outbuildings and comments on the immediate locality.*

- *A thorough analysis of the structure of the property including the composition of the chimney stacks, roofs, external and internal walls, partitions and flooring (where accessible).*

- *A complete inspection of the remaining elements of the building and general commentary on the state of repair. Items inspected will include flashings, parapets, gutters, downpipes, roof spaces, damp-proof courses, external and internal joinery (e.g. window frames and staircases etc) and fireplaces.*

- *General checks should be carried out to identify defects such as dampness, beetle decay, dry rot or any other timber defects. Any cracking to the structure should also be noted and analysed.*

- *A visual inspection of the services installed in the property namely, electricity, gas, water (including the general condition of the plumbing and sanitary fittings), central heating system and any drainage facilities.*

- *Most structural surveys will have a summary of necessary repairs and their approximate cost. The surveyor should make a reflective analysis of his observations and conclude with some recommendations.*

- *Finally, if the surveyor has grounds to suspect that a serious defect may exist which requires further investigation by a specialist he should recommend a report by an independent expert.*

The Home Buyers' Survey & Valuation Report

It should be clear by now that valuation reports are quite different from structural survey reports, both in their objectives and the detail which they claim to cover.

In the light of widespread misconception by purchasers as to what a valuation report actually represents, the RICS intervened by introducing a series of 'survey and valuation' reports. Initially, there were two separate report forms for house and flat purchases but these were combined in August 1993 to form the Home Buyers' Survey & Valuation Report. Surveyors are not obliged to use this new report.

This report has been met with a mixed reception. Some mortgage lenders have insisted that their valuation surveyors use the report, but the general consensus amongst practising surveyors is that it is too detailed in some parts and insufficient in others.

The main criticism is that the information required by the form goes beyond what is reasonably expected from a valuation survey. Many surveyors feel that in view of the modest fee chargeable for a mere valuation survey, they could not justify spending the extra time needed in order to complete the Home Buyers' report.

Chapter Three

The Duty of Care Owed by Surveyors

Chapter One gave an overview of the basic principles of negligence and their development in English law. This chapter will now look at the common law duty of care in relation to the liability of the surveyor and how the case law has developed this concept.

The early authorities in the field of professional negligence, which were considered in Chapter One, predominantly focus on the negligent actions of the defendant. The primary difference, when considering surveyors' negligence, is that the alleged act of negligence almost always takes the form of incorrect or inaccurate advice given in the form of a written or typed report.

Negligent misstatement and pure economic loss

The concepts of 'negligent misstatement' and the resulting 'pure economic loss' overlap to some extent with what has already been discussed in chapter one but they are of particular relevance to professional negligence and even more important in the field of surveyors' negligence. The landmark case of **Hedley Byrne & Co. v. Heller & Partners [1964]**[31] had the effect of extending the concept of duty of care so that it applied to negligent words or statements and not merely negligent acts, hence liability for negligent misstatements was established as a principle of law.

In **Hedley Byrne** a misleading credit appraisal report was compiled by the defendants, who were the bankers of a third party company, Easipower Ltd. Hedley Byrne relied on the favourable report, Easipower subsequently went into liquidation and Hedley Byrne suffered loss as a result. Lord Devlin stated, *obiter,* that there was "a duty to take care in word as well as in deed" and that where the relationship between the parties was "equivalent to contract" there was an assumption of responsibility. Therefore, people holding themselves out as being skilled advisers could be liable in negligence to other persons who rely upon their advice.

[31] [1964] AC 465

Hedley Byrne was decided in favour of the defendants, due primarily to the presence of an express disclaimer which had been incorporated into the report. Nevertheless, this case is of importance in relation to the issue of pure economic loss. In tort, economic loss is usually recoverable where the loss is 'consequential on some form of physical damage'.[32] However, where there is 'pure' economic loss, which can be described as purely financial loss which neither stems from physical damage nor personal injury, the courts are more reluctant to award damages in order to compensate claimants.[33]

The rationale behind this principle is the fear that recovery of economic loss in the tort of negligence will give rise to the spectre of liability in an indeterminate amount for an indeterminate time to an indeterminate class.[34] It was this view which prompted the House of Lords to come to the

[32] Cases & Commentary on Tort, Harvey & Marston, p.42
[33] Lord Denning MR in Spartan Steel & Alloys Ltd v. Martin & Co [1972] 3 All ER 557
[34] per Cardozo CJ in Ultramares Corporation v. Touche (1931) 255 NY 170; 174 NE 441

restrictive decision in **Caparo Industries plc v. Dickman [1990]**[35]. In this case it was held that liability for economic loss caused by negligent misstatements was limited by the burden of having to prove that the defendant knew the inaccurate information would be relied on by the plaintiff either as an individual or as a member of an identifiable class.

However, an earlier case illustrates how the concept of pure economic loss has been approached in the area of surveyors' negligence. In **Yianni v. Edwin Evans [1982]**[36] evidence was put forward to show that less than 15 per cent of the public who purchased property actually had an independent survey carried out over and above the valuation report carried out by order of the mortgage lender. It was held that the surveyor was liable for the economic loss suffered by the plaintiff as this would not expose him to unlimited liability (which was the fear in **Caparo**), since liability would be limited to the purchaser named in the lender's instructions.

---

[35] [1990] 2 AC 605; [1990] 1 All ER 568
[36] [1982] 1 QB 438

The decision in Smith and Harris

A landmark ruling in the area of surveyors' negligence was given in the joint House of Lords' cases of **Smith v. Eric S Bush** and **Harris v. Wyre Forest District Council [1989]**[37]. The House of Lords sought to define the criteria for establishing when a surveyor owes a duty of care for a negligent misstatement. As with the case of **Yianni**, in **Smith** and **Harris** the valuation survey report in question had been prepared for a mortgage lender rather than the property purchasers who, in these cases, were bringing the actions. However, their Lordships were in agreement that **Yianni** had been correctly decided and Lord Griffiths gave the following ruling:

> *'A valuer who values a house for a building society or a local authority for the purposes of a mortgage application for a typical house purchase knowing that the mortgagee will probably, and the mortgagor will*

[37] [1989] 2 All ER 514

certainly, rely on the valuation, and knowing that the mortgagee is an intending purchaser of the house and has paid for the valuation (even if indirectly through administration fees etc.), owes a duty of care to both parties to carry out his valuation with reasonable skill and care.'

Once again, as in the case of **Yianni**, Lord Griffiths explained that the extent of such liability was limited to the prospective purchaser of the house and did not extend to subsequent purchasers. Furthermore, it was held that even if a building society or local authority are under a statutory duty to value the house themselves, this would not negate the surveyor's common law duty of care to the purchaser. The only unanswered question left by **Smith** and **Harris** is whether the same duty of care would be owed to the purchasers of a more expensive property.

This uncertainty arises as a direct result of Lord Griffiths' judgement in which he expressly reserved his position in relation to whether the same scope for duty of care would arise in the case of a valuation of industrial property or expensive houses. Lord Griffiths stated that "In such cases it may well be that the general expectation of the behaviour of the purchaser is quite different. With very large sums of money at stake, prudence would seem to demand that the purchaser obtain his own structural survey."

As a result of Lord Griffiths' judgement there remains some confusion as to when a purchaser will be owed a duty of care. Traditionally, questions regarding the surveyor's common law duty of care would only have caused a problem when the purchaser had no contractual relationship with the surveyor (i.e. when the surveyor is instructed by the mortgage lender). A statutory equivalent to the common law duty of care, is implied into contracts for the supply of services, pursuant to section 13 of the **Supply of Goods and Services Act 1982**. The implied term stipulates that the party supplying the service should do so with reasonable care and skill.

The Contracts (Rights of Third Parties) Act 1999

The much criticised[38] doctrine of privity of contract has always prevented a purchaser from suing under a contract entered into between his mortgage lender and the surveyor. Having said this, the recent enactment of the **Contracts (Rights of Third Parties) Act 1999**, which applies to contracts entered into on or after 11th May 2000, may, in specific circumstances, enable purchasers, who are not privy to such contracts, to enforce its terms.

The difficulties that a purchaser is likely to face in bringing an action in accordance with the new Act are twofold.

First, section 1(1) of the Act is clear that a third party will only be able to enforce a term of a contract where the contract expressly provides that a third party may do so, or where the term confers a benefit on the third party who wishes to enforce it.

[38] see Swain v. The Law Society [1983] 1 AC 598 at 611 per Lord Diplock

On this basis it is doubtful whether terms implied by statute will be enforceable by third parties.

Secondly, section 1(3) of the Act stipulates that the third party wishing to enforce a term of a contract must be either named in the contract, identified as a member of a class or answer a particular description.

It is highiy likely that prudent mortgage lenders will amend their standard instructions to valuation surveyors so as to incorporate an express exclusion of the rights conferred under the new Act. However, the full implications of this new legislation will undoubtedly unfold in due course as and when purchasers seek to rely on the Act's provisions.

As will have been gleaned from this chapter, the proposition that surveyors owe their clients a common law duty of care is now considered to be trite law. However, the same certainty does not seem to apply to other important aspects of claims against negligent surveyors, namely the required 'standard of care' and the 'measure of damages', both of which have proven to be far more problematical. Accordingly, Chapters Four and Five deal with each of these legal issues respectively.

Chapter Four

The Standard of Care Required of Surveyors

There is often some confusion on the part of house purchasers (and amongst surveyors it would sometimes appear) as to what a valuation survey purports to inspect. This lack of understanding is frequently the primary cause of litigation involving allegations of surveyors' negligence. A number of recent court decisions have attempted to draw a distinction, in terms of standard of care, between the two types of survey.

The majority of authorities discussed in this chapter focus on the most common causes of actions against surveyors: first, where the surveyor has failed to identify fundamental defects in the property which he has surveyed; and secondly, cases in which the defendant surveyor has over-estimated the value of the property in question.

The second of these two categories of case law has led to many decisions of both the Court of Appeal and the House of Lords. The development of a principle known as the 'margin of error' has been an inevitable progression of the law in this area mainly as a result of the volatile nature of the property market during the late 1980's and early 1990's. This aspect of the surveyor's expected standard of care is of the utmost importance to residential and commercial valuation surveyors alike.

Distinguishing between Structural Surveys and Valuations

It should be evident by this stage that the majority of claims against surveyors are based on tort rather than claims for breach of contract, not least because of the lack of a contractual relationship between the parties in most cases. However, the standard of care that may be expected from a surveyor is the same irrespective of whether or not a cause of action is the breach of a contract, the breach of a tortious duty of care, or even both.

It is necessary though, as the courts have realised, to distinguish between the two types of survey when considering the appropriate standard of care. This distinction was emphasised in the case of

Gibbs v. Arnold Son & Hockley [1988][39] where the court held "there is a world of difference between a valuation survey and a structural survey." and suggested that "defects which may be important for the latter may be relatively unimportant for the former."

In the case of **Roberts v. J Hampson & Co [1988]**[40] it was stressed that when dealing with a mortgage valuation survey "it is understood that [the surveyor] is making a limited appraisal only. It is, however, an appraisal by a skilled professional man"[41] This analysis of the valuation survey was approved by the House of Lords in the joint cases of **Smith v. Eric S Bush** and **Harris v. Wyre Forest District Council [1989]**[42] in which Lord Templeman was also cautious not to impose too high a standard of care on valuation surveyors.

However, some commentators still feel that many lower court judges have persisted in failing to grasp the quite considerable

---

[39] [1989] 2 EGLR 154
[40] [1988] 2 EGLR 181
[41] see Jackson & Powell on Professional Negligence pp.211 for Kennedy J's full quote
[42] [1989] 2 All ER 514

difference between a structural survey and a mere valuation with regard to applying the appropriate standard of care[43]. A case which lent some support to such criticism is **Sneesby v. Goldings (a firm) [1995]**[44] but that decision was overturned by the Court of Appeal which ruled that the first instance judge had placed too high a duty on a valuation surveyor by "blurring" the distinction between structural surveys and mere mortgage valuations.

It would appear that, generally, fears concerning the judiciary's competence in dealing with issues of standard of care are unfounded. Indeed, in the case of **Assured Advances Ltd v. Ashbee & Co [1994]**[45] Edwards J. clearly exhibited a realistic understanding of the distinction and its implication on the standard of care issue. He stated, having noted that the defendants were only paid a fee of £50, that "nothing more than a quick market valuation is looked for and that such an assessment would not involve a survey."

[43] see 'Lenders and valuers in court again' by John Murdoch LLB, Estates Gazette Jan 21, 1995
[44] [1995] EGCS 201
[45] [1994] EGCS 169

Exercising reasonable skill and care

Regardless of whether a surveyor is carrying out a valuation survey or a structural survey, certain standards will, in all cases, be required of him if he is to evade liability for professional negligence. These standards can be viewed as the accepted practices of the surveying profession and guidance notes published by the RICS which are mentioned earlier assist in outlining many of these, but how has the case law in this area interpreted the standard of care on a factual basis?

The Bolam test, as explained in chapter one, has direct applicability to the area of surveyors' negligence and thus it can be said that a surveyor will be judged against the standard of what others in his profession would reasonably have done in the same situation. It is worthy to note however, that following a 'common practice' will not necessarily provide conclusive evidence that a surveyor has not been negligent, as was the ruling in an early surveyors negligence case, **Last v. Post**[46].

---

[46] (1962) 159 EG 240

Therefore, it is true to say that a surveyor must exercise all the relevant skill and care that may be expected from such a surveyor.[47] Indeed, in **Smith v. Eric S Bush** and **Harris v. Wyre Forest District Council [1989]**[48], Lord Templeman, in his judgement, acknowledged that valuation surveyors must exercise reasonable skill and care by taking into account "major defects which are or ought to be obvious to him in the course of a visual inspection."

The valuation surveyor in the case of **Bere and Parchment v. Slades [1990]**[49] was held to have used reasonable skill and care despite the fact that he had failed to warn the plaintiffs of the existence of a defect in external cellar walls. However, the court explained that this was mainly due to the presence of two factors. Firstly that the defendant had produced "a thoroughly competent report, carefully done and masterly succinct.", and secondly that the damage complained of by the plaintiffs was not sufficient to

---

[47] taken from 'Professional Negligence & Insurance Law - An Industry Report' p.25

[48] [1989] 2 All ER 514

[49] [1989] 2 EGLR 160

create a duty on the part of the defendant to inform the plaintiffs of the need for a further investigation.

In contrast, the Court of Appeal held in **Ezekiel v. McDade [1994]**[50] that the surveyor ought to have noticed a gap in the loft caused by the displacement of a concrete beam or 'purlin'. Sir John Megaw stated that the defendant had failed to use reasonable skill and care by only carrying out a "head and shoulders" inspection of the roof space. Therefore, it is expected that a surveyor (on a structural survey at least) will inspect thoroughly, every area which is physically accessible.

The principle of 'follow the trail'

Another difficulty for defendant surveyors seems to arise where signs of a defect are evident during the inspection but are not acted upon. The important case of **Roberts v. J Hampson & Co [1988]**[51] established a principle which has since become an integral element of the standard of care required of the surveyor. If the surveyor has specific grounds for suspicion and the trail of suspicion requires the

[50] [1994] EGCS 194
[51] [1988] 2 EGLR 181

surveyor to either move furniture or lift carpets, then the valuer must take all reasonable steps to 'follow the trail' until he or she has all the information which he or she reasonably requires to make a sufficiently accurate report or valuation.

In some instances the surveyor may be physically barred from establishing the extent or even existence of a suspected defect. Many occupied properties have fitted carpets and heavy furniture and hence following the trail can often prove somewhat difficult. When this is the case, a prudent surveyor will usually note the possibility of a defect and suggest an appropriate course of action, whether it be a further, more specialised inspection or some other precautionary measure.

For example, in **Eley v. King & Chasemoor (a firm) [1989]**[52], the surveyor discovered some slight cracking to the internal walls during a structural survey but was unable to lift any floor boards so as to inspect the foundations. Suspecting the presence of either 'subsidence', 'ground heave', 'settlement' or 'landslip' the surveyor advised

---

[52] [1989] NLJ 791

the purchasers to seek insurance in order to cover themselves for such an eventuality. The Court of Appeal held that on these facts the surveyor was not negligent because insurance provided the best means of covering the risk.

Valuation surveyors or economists?

Where a surveyor has been instructed to carry out a valuation survey of a property, his or her primary job is to assess the open market value of the property in question at the date of valuation. It is therefore implicit that the surveyor has considerable knowledge of property prices. For this reason, many practising surveyors will only undertake survey work in areas which are familiar to them. This is because property markets can vary dramatically due to various economic factors and thus sometimes, even areas relatively local to one another may experience significant disparities.

A prudent valuation surveyor will inevitably seek the assistance of a local estate agent in order to ascertain a general price range for the property in question. However, it is obviously the responsibility of the surveyor to arrive at a final value once he has assessed the overall condition of the property. A substantially inaccurate valuation, which no other reasonably competent valuer could have arrived at, will almost certainly constitute professional negligence and thus expose the surveyor to the liability for any foreseeable loss.

Difficulties have arisen in recent case law where economic instability on a larger scale has resulted in negligent valuations. During the early 1990's the economic recession in this country had a drastic effect on the property market. It was this economic climate which resulted in the collapse of property prices and subsequently, the inevitable spate of negligence claims against unsuspecting valuation surveyors. Of these cases, two of the first to come to court were **United Bank of Kuwait v. Prudential Property Services Ltd [1994]**[53] and **Banque Bruxelles Lambert SA v. Eagle Star Insurance Ltd [1994]**[54].

It should be noted that both of these actions were brought by commercial lenders against the

---

[53] [1994] 2 EGLR 100
[54] [1994] 2 EGLR 108

defendant surveyors rather than by the purchasers of the properties. These cases serve as a good illustration of the initial judicial inconsistency that prevailed in the early part of 1994.

In the first of these cases, **United Bank of Kuwait**, the mortgage lenders argued successfully that, 'but for' the valuation surveyor's negligence, there would have been no mortgage loan granted at all. Therefore, it was held by Gage J that the surveyor's standard of care did extend to fluctuations in property prices brought about by economic factors.

No less than two weeks later the ruling in **Banque Bruxelles Lambert** appeared to blatantly ignore the decision in **United Bank of Kuwait**. According to Phillips J, since the lender did not rely on the valuer for protection against movements in the property market, it would not be fair and reasonable to make the valuer carry that risk. Subsequently, the approach in **Banque Bruxelles Lambert** was affirmed as the correct view when the High Court gave its ruling in the case of

Nyckeln Finance Co Ltd v. Stumpbrook Continuation Ltd [1994][55].

After **Nyckeln Finance**, a flood of cases including **BNP Mortgages Ltd v. Goadsby & Harding Ltd [1994]**[56] and **Assured Advances Ltd v. Ashbee & Co [1994]**[57] supported the view that it is the lender, and not the valuer, who bears the risk that the market may turn.

In **Assured Advances** the evidence given included relevant indices showing that property prices at that time were still rising. On the strength of this evidence the judge rejected the plaintiff's argument that the first early warning signs of the market collapse were already apparent.

The conflict on this point went to the Court of Appeal as a joint appeal of six first instance decisions, where it was held that such losses were recoverable against negligent valuers.

---

[55] [1994] 33 EG 93
[56] [1994] 42 EG 150
[57] [1994] EGCS 169

A further appeal was made to the House of Lords in **South Australia Asset Management Corp v. York Montague Ltd [1997]**[58] and is examined in detail in Chapter Five. The Lords unanimously held that losses attributable to a collapse in the property market were mot recoverable – the standard of care extended only to the direct consequences of the surveyor's advice being wrong.

The 'margin of error' approach

The judgement in **Assured Advances** went further in that the court considered whether a particular valuation would merely "raise eyebrows a little" or whether it was truly negligent. Judge Gareth Edwards QC stated that it was permissible to find a central accurate valuation on the basis of the expert evidence adduced and to establish a percentage above and below it, anything outside of which would be a *prima facie* negligent valuation.

This method for determining whether a defendant surveyor's valuation falls short of the required standard of care by arriving at a figure which is

58 [1997] AC 191; [1996] 2 EGLR 93

outside the range of any competent surveyor, is often referred to as the 'margin of error' approach. The use of margins of error is now common place in negligent valuation cases and 'brackets' of between 10% and 20% appear to be the normal range.

In **The Mortgage Corporation v. Halifax (SW) Ltd and Anscombe & Ringland Ltd [1999]**[59] the first and second defendants were both engaged by the lender to each carry out a valuation on a large residential property. The first defendant valued the property at £5.25 million and the second defendant arrived at a figure of £5 million. On hearing the expert evidence of each party the judge found that the property was in fact worth £4.15 million and that the appropriate margin of error in the circumstances was a bracket of 15.6%.

On this basis, £4.8 million was the highest non-negligent valuation that could have been given by either defendant. Accordingly, both defendants had given valuations that were outside the margin of error and thus both were liable.

---

[59] [1999] Lloyd's Rep. PN 159

A question which has arisen since **Assured Advances** is whether a valuation which does fall within the accepted bracket is still capable, in certain circumstances, of being negligent. In **Merivale Moore plc v. Strutt & Parker [1999]**[60] the Court of Appeal took the view that so long as the given valuation falls within the bracket, which has been accepted by the court as the appropriate margin of error, the valuer will not be liable.

However, in **Arab Bank Plc v. John D Wood Commercial Ltd. and Weatherall, Green & Smith [2000]**[61], Mance L.J. suggested in *obiter* that there was no reason why a valuer, who is shown to have made a discrete and clearly identifiable error that feeds through pound for pound into the bottom line of the valuation, should not be found liable for negligence, even though the bottom line figure would otherwise lie within an acceptable margin of error.

---

[60] [1999] 2 EGLR 171

[61] [2000] 1 WLR 857; [1999] EGCS 133

When should a valuer discount the valuation figure

In **UCB Corporate Services Ltd v. Halifax (SW) Ltd and Keith Ralphs [2000]**[62] the claimant instructed the second defendant to give a valuation of commercial property on the basis of a 90-day sale in the event of the claimant taking possession as mortgagee.

A valuation of £665,000 was given but following the borrower's default the property was repossessed and recovered only £325,000 on the sale. The first instance judge found that the true value had been £644,000 and that the appropriate margin of error was 10%. Therefore the valuation given had not been negligent and the claimant's case was accordingly dismissed.

On appeal to the Court of Appeal, the claimant contended that the judge had erred in not discounting his finding of the property's value at £644,000 in order to reflect the fact that the valuation sought was a 90-day sale value.

It was held that the question of whether a discount should have

[62] [2000] 16 EG 137

been made to the valuation depended upon the market conditions at that time. The trial judge had been entitled to rely on expert evidence that it was reasonable for the second defendant not to have made a discount in the circumstances. The court stressed that there is no irrebuttable presumption that a forced 90-day sale value will be less than that of the open market value.

Chapter Five

Assessing the Appropriate Measure of Damages

Assuming that a defendant surveyor has been found to have compiled a negligent survey report, the court then has to consider the question of an appropriate remedy. With any form of litigation it is always vital that the claimant's advisor knows from the outset what remedy his or her client is seeking and whether it is available at the conclusion of legal proceedings.

Therefore, it is prudent to examine the remedy or remedies to which a successful claimant may be entitled, especially as there has been some conflicting case law in this particular area of surveyors' negligence. It is fair to say that claimants who pursue actions against negligent surveyors are seeking compensation by an award of damages.

The question of finding the appropriate measure of damages is a contentious issue but general principles for the awarding of damages are important here. In tort, the rationale behind the awarding of damages is to put the claimant back into the position he would have been in, had the tort not occurred. This can clearly be contrasted with the principle of contractual damages which aims to put the claimant forward in time, into the financial position they would have been in, had the contract been performed correctly and fully.

Two categories for the assessment of damages

The case law in this area can be divided into two categories:-

First are the cases which focus on the appropriate measure of damages for compensating mortgage lenders who have approved loans on the basis of negligent valuations.

'Transaction Cases' and 'No-Transaction' cases

In such cases, there has been a tendency to distinguish between 'transaction' cases where the lender would have reduced the loan if the true value had been known, and 'no-transaction' cases where the loan would have been refused completely.[63] The House of Lords has subsequently ruled that such an

[63] see 'Negligent valuers: falling markets' in Legal Notes, Estates Gazette, Sept 17 1994

approach is not based on principle and should be abandoned.[64]

Secondly, there are the cases involving the assessment of damages which are appropriate for compensating purchasers of property who subsequently discover defects which were missed by the surveyor's report.

Claims by lenders in respect of valuations

With regard to the former of these two categories, one of the earliest surveyors' negligence cases which raised the issue of assessment of damages was the case of **Baxter v. F W Gapp & Co Ltd [1939]**[65]. In this case the plaintiff, who was the mortgage lender, argued that they would not have entered into the transaction at all had the surveyor valued the property more carefully. On the question of damages the judge considered that the plaintiff was entitled to recover the whole loss suffered as a result of reliance on the surveyor's over-valuation of the property. On the strength of this

---

[64] see South Australian Asset Management Corp v. York Montague Ltd [1997] AC 191

[65] [1939] 2 KB 271; [1939] 2 All ER 752

principle, **Baxter** was followed for many years and cited often.[66]

However, the more recent case of **Swingcastle & Co v. Gibson [1991]**[67] gives an example of the way in which the courts have chosen to approach the assessment of damages in relation to negligent mortgage valuations. The facts were that the surveyor over-valued the property at £18,000, which resulted in the plaintiff lending the purchaser a sum of £10,000. When the new owners subsequently defaulted on the mortgage, the plaintiff repossessed the property and was only able to recover a forced sale price of £12,000. Accordingly, Swingcastle brought an action against the surveyor, claiming for their loss.

House of Lords ruling loss of the unpaid interest cannot not be recovered

At first instance, it was held that the measure of damages to which the lender was entitled, was the whole of the loss sustained including the principal sum advanced and the unpaid interest which the plaintiff stood to make from the mortgagors.

---

[66] see London & South of England Building Society v. Stone [1983]

[67] [1991] 2 AC 223; [1991] 1 EGLR 157

The surveyor appealed against this decision but the Court of Appeal held that they were bound by **Baxter** and thus the appeal was dismissed. However, the House of Lords did allow a further appeal by ruling that the loss of the unpaid interest could not be recovered. Such damages could only have been recovered if the valuer had given a warranty that his valuation was accurate. This was not so and the case had been brought solely on the basis of breach of duty of care.

The decision in South Australia Asset Management v. York Montague Ltd [1997]

The subsequent House of Lords decision in **South Australia Asset Management Corporation v. York Montague Ltd [1997]**[68], which incorporated three appeals from the Court of Appeal, further clarified the law in relation to the assessment of damages in lender-valuer cases and now provides a comprehensive statement of the law on this point.

The defendants overvalued a development site in London's

[68] [1997] AC 191; [1996] 2 EGLR 93

Docklands at £15 million on 29th June 1990 and accordingly loaned £11 million to the purchaser. At first instance[69] May J. found that, on the basis of the expert evidence, the property had been worth £5 million but that due to further property price decline only £2,477,000 was recovered. The trial judge found in favour of the plaintiff (less 25% of the amount claimed to reflect contributory negligence) that the whole of the loss sustained, some £9.75 million was within the scope of the valuers' duty.

United Bank of Kuwait v. Prudential Property Services Ltd [1995] and Nykredit Mortgage Bank v. Edward Erdman Group Ltd [1995]

Similar facts and rulings were given in the other two cases being appealed, namely **United Bank of Kuwait plc v. Prudential Property Services Ltd [1995]** and **Nykredit Mortgage Bank plc v. Edward Erdman Group Ltd [1995]**[70]. It was clear that with the substantial number of inaccurate high-value commercial

[69] [1995] 2 EGLR 219
[70] [1995] 1 EGLR 129

valuations which had occurred during the property market collapse of the early 1990's, a decisive authority was needed to define the extent of the loss that a negligent surveyor would be liable for.

In the appeal to the House of Lords, the defendant surveyors all contended that they should not be liable for the decline in the property market which had resulted in values falling during the period between the original valuation and the realisation of the security by the respective lenders.

The lenders asserted that had the true value of the properties been known, the loan advances would not have been made and thus the lenders should be entitled to the difference between the amount lent with interest and the net sum recovered.

It was finally held by Lord Hoffman, who gave the leading judgement, that the correct measure of damages is the loss suffered by the plaintiff, which is attributable to the inaccuracy of the valuation, upon the plaintiff agreeing to advance the loan on the assumption that the valuation advice was correct.

Accordingly, the recoverable damages in such cases must be limited only to the foreseeable consequences of the advice being wrong, which Lord Hoffman explained was the difference between the actual valuation and what would have been a correct valuation at that time. This principle has become commonly referred to as the 'SAAMCO cap'.

The Mortgage Corporation v. Halifax (SW) Ltd and Anscombe & Ringland Ltd [1999]

As explained in Chapter Four, both defendants were found to be negligent in **The Mortgage Corporation v. Halifax (SW) Ltd and Anscombe & Ringland Ltd [1999]**[71]. However, because the first defendant gave a valuation that was £250,000 higher than the second defendant, the recoverable loss was potentially different against each defendant. This issue was resolved by the judge ruling that where two valuers were jointly and severally liable, only one 'cap' is applicable to the claims against each defendant, that being the loss on the lower valuation.

[71] [1999] Lloyd's Rep. PN 159

Claims by purchasers in respect of missed defects

The second category (as explained above) has also been the focus of much debate. In cases where the surveyor is being sued by the purchaser of a property on the basis of a negligent survey report, the main issue for the courts has been whether the purchaser should recover the 'cost of cure', otherwise known as the 'cost of repairs', or the 'diminution in value' of the property.

This distinction is significant by virtue of the fact that the cost of repair will invariably exceed the diminution in value. This is because the latter is an assessment of the reduction in sale price that would have been negotiated between the vendor and purchaser had the defect been discovered, rather than the latent amount that the vendor would actually need to spend in order to remedy the defect in its present state.

The diminution in value of a property is ordinarily ascertained on the basis of one or more retrospective valuations by expert witnesses. The inevitable battle of the experts which often ensues in such cases is likely to be eliminated to some extent by Part 35.7 of the Civil Procedure Rules which allows the court to direct that a single joint expert must be used.

'Cost of cure' or 'Diminution in value'

British Westinghouse Electric & Manufacturing Co Ltd v. Underground Electric Railways Co [1912]

In the case of **British Westinghouse Electric & Manufacturing Co Ltd v. Underground Electric Railways Co [1912]**[72] Lord Haldane LC was of the opinion that "the proper measure of damages is the difference between the value in its assumed good condition and the value in the bad condition which should have been reported to the client".

Lord Haldane qualified this view by explaining that had the plaintiff received a proper report "[he] would either have refused to have anything to do with the house, in which case he would have suffered no damage, or he would have bought it for a sum which represented its fair value in bad condition."

---

[72] [1912] AC 673

British Westinghouse became regarded as settled law and thus a long line of case law followed its rationale in relation to the appropriate measure of damages. In **Philips v. Ward**[73] the Official Referee awarded the difference in value of the property as described by the negligent surveyor rather than the cost of repairs.

Dodd Properties (Kent) Ltd v. Canterbury City Council [1980]

Similarly, the case of **Dodd Properties (Kent) Ltd v. Canterbury City Council [1980]**[74] held that the buyer is not entitled to remedy defects at the expense of the survey. Instead, "damages are to be assessed at the time of the breach according to the difference in price the buyer would have paid if the report had been carefully made."

County Personnel Employment Agency Ltd v. Alan R Pulver & Co [1987]

Indeed, the Court of Appeal went one step further when it was suggested in the case of **County Personnel Employment Agency Ltd v.**

---

[73] [1956] 1 WLR 471; [1956] 1 All ER 874
[74] [1980] 1 WLR 433; [1980] 1 All ER 928

Alan R Pulver & Co [1987][75] by Bingham LJ that "on the authorities as they stand the diminution in value rule appears almost always, if not always, to be appropriate where property is acquired following negligent advice by surveyors." However Bingham LJ did acknowledge that this was not an invariable approach and that there may be cases in which the presence of special circumstances make the diminution of value an inappropriate measure of damages.

Hipkins v. Jack Cotton Partnership [1989]

Subsequently, Judge Scott Baker failed to apply the diminution in value rule in **Hipkins v. Jack Cotton Partnership [1989]**[76] some two years later. It appeared that on the facts there was nothing to distinguish this case from those which had gone before and neither was there a special circumstance to warrant this departure. Therefore the damages awarded to cover the cost of repairs were unjustified. Nevertheless, this case started a judicial trend which saw a move away from the traditional

[75] [1987] 1 WLR 916; [1986] 2 EGLR 246
[76] [1989] 2 EGLR 157

diminution in value rule towards the cost of repairs assessment.

Following the decision in **Hipkins**, a couple of articles in the Estates Gazette focused on Lord Justice Bingham's judgement in **County Personnel**.[77] In these articles the views of Judge Fox-Andrews QC were contrasted with those of Judge Bowsher QC.

Judge Fox-Andrews was quoted as saying "the difference in value or, as preferably described, overpayment is not invariably the appropriate measure of damage." However, the learned judge did also state that "in the absence of 'special circumstances' the rule ought to be invariable." Furthermore, when pressed, Judge Fox-Andrews felt that evidence suggesting that the purchaser would not have bought the property had he known, did not constitute a special circumstance in that sense.

On the other hand, the article also pointed out that Judge Bowsher, another Official Referee, in the course of one of his judgements, had clearly regarded it as a 'special

[77] see 'Negligent surveys - the measure of loss' December 8, 1990

circumstance' that the plaintiff would not have purchased the property had he known the truth. On this premise, the learned judge was of the opinion that such a plaintiff should be compensated by his negligent surveyor for all the repair costs which had arisen as a result of the undetected defects.

Syrett v. Carr & Neave (a firm) [1990]

This view was subsequently reflected in Judge Bowsher's ruling in the case of **Syrett v. Carr & Neave (a firm) [1990]**[78]. The facts of this case were that the plaintiff paid £300,000 for a house in the New Forest following an encouraging, albeit negligent, structural survey report which was compiled by the defendant surveyors. Subsequently, it was found that the property had a severe death-watch beetle infestation, the eradication of which was estimated at £78,000.

The defendants ultimately conceded that they had been negligent but still argued that the damages payable to Syrett should be limited to the difference in value between the property as described and in its true state, which was calculated

[78] [1990] 2 EGLR 161; [1990] 6 Const. LJ 305

as £55,000 at the date of purchase. Judge Bowsher rejected the defendant's argument and held that the plaintiff was entitled to the cost of the repairs.

The decision in Watts v. Morrow

Judge Bowsher QC was also the first instance judge in what transpired to be a landmark case in this area of professional negligence. Not surprisingly, in **Watts v. Morrow [1991]**[79] Judge Bowsher stuck to his guns and held once again that the correct measure of damages was the cost of repairs, making particular reference to his recent decision (at that time) in **Syrett.** However, the defendant surveyors in this case appealed against the decision in terms of what damages should be recoverable and it was the ruling of the Court of Appeal which makes **Watts v. Morrow** a landmark case.

Watts v. Morrow Appeal Court Decision

The facts of Watts v. Morrow were that the plaintiffs, who were well paid City executives sought to purchase a second 'country' home in rural Dorset. They instructed their own independent structural survey to

[79] [1991] 1 WLR 1421; [1991] 2 EGLR 152

be done by the defendant on an 18th century listed property which they were interested in buying. The survey report outlined some minor defects but failed to note several more serious faults, including the fact that the roof needed renewing. The plaintiffs took it upon themselves to repair these defects and as a result incurred total costs of £33,961.

After reviewing the facts of the case, the Court of Appeal overturned the first instance decision of Judge Bowsher QC in favour of the 'diminution in value' approach. In his leading judgment, Ralph Gibson LJ made an analytical examination of the assessment of damages in cases such as these. The main points that were raised are summarised below:

Guidelines for the assessment of damages

- In cases involving the purchase of private property (as opposed to commercial) the *prima facie* rule for assessing damages as in **Philips v. Ward** is correct. Namely, the difference between the value of the property in the condition described by the negligent report and the value as it should have been described.

- There is no basis for awarding the cost of repairs because the surveyor gives no warranty that the purchaser will not incur any repair costs.

- The *prima facie* rule may not apply in some circumstances such as where the purchaser extricates himself by re-selling. In such a case the damages recoverable may not necessarily be limited to the difference of value.

- Proof that the plaintiff, properly advised, would not have gone ahead and purchased the property does not make the difference in value rule inapplicable.

- **Hipkins v. Jack Cotton Partnership** and **Syrett v. Carr & Neave** were both wrongly decided. Additionally, the plaintiff's decision to carry out repairs does not alter the proper measure of damages.

- Having stated the above, the Court of Appeal held that the £33,961 in original damages was to be reduced to £15,000 so as to reflect the diminution in value.

Following the ruling of Ralph Gibson LJ in the Court of Appeal, Mr and Mrs Watts sought leave to appeal the decision in the House of Lords. However, leave to appeal was refused and thus **Watts v. Morrow** can now be regarded as settled law.

Patel v. Hooper & Jackson [1998
Attempt to revert to the cost of repair approach.

Since the Court of Appeal's ruling in **Watts v. Morrow**, there have been occasional attempts by trial judges to revert to the cost of repair approach. Most recently, in the case of **Patel v. Hooper & Jackson [1998]**[80] the trial judge did award the Claimant purchaser an amount of damages in the sum of £25,000, which reflected the diminution in value of the property.

However, in addition the judge also awarded damages to compensate the Claimant for the costs of his mortgage interest

---

[80] [1998] EGCS 160

payments, endowment policy payments and estate agents' and solicitors' fees on the basis that the Claimant would not have purchased the house had the survey revealed the property's true condition.

On appeal, the Court of Appeal unsurprisingly overturned the decision of the trial judge, ruling that the sole award of £25,000, which reflected the property's diminution in value, was the correct assessment of the Claimant's recoverable loss.

Damages for 'distress and inconvenience'
Perry v. Sidney Phillips & Son [1982]

A further issue which can arise in relation to assessing damages is the remoteness of damage argument. When dealing with claims against negligent surveyors it should be realised that additional damage, less obvious than the type of loss considered above, may have been incurred by the plaintiff. The predominant type of 'alternative' loss, which has been considered in several cases, has been labelled as damages for 'distress and inconvenience'. The rationale behind this head of damage is the need to compensate the plaintiff for the physical and/or psychological effects of living for prolonged periods in a defective property or having to endure extensive repair work.

The first significant case that addressed the issue of damages for distress and inconvenience was **Perry v. Sidney Phillips & Son [1982]**[81]. Lord Denning, who was Master of the Rolls at that time, held that damages for distress and inconvenience should be kept to a modest sum. This was the first sign of a general judicial reluctance to award over-inflated damages which would have the effect of punishing or penalising the defendant surveyors.

Hayes v. Dodd & Charles (a firm) [1990]

In the case of **Hayes v. Dodd & Charles (a firm) [1990]**[82], Staughton LJ, giving judgment in the Court of Appeal, held that damages for distress and inconvenience were too remote and therefore not recoverable. Staughton LJ continued by saying that, as a matter of policy, damages for mental distress are limited to a class of case "where the contract which has been broken was itself a

[81] [1982] 1 WLR 1297; [1982] 3 All ER 705
[82] [1990] 2 All ER 815

contract to provide peace of mind or freedom from distress."

However, once again the landmark case of **Watts v. Morrow [1991]**[83] seems to have shaped the law in this aspect of surveyors negligence. At first instance, Judge Bowsher QC had awarded Mr & Mrs Watts £4,000 each on the basis of distress and inconvenience.

As expected the Court of Appeal felt that this figure was much too large and thus reduced the award to £750 each. Ralph Gibson LJ explained that the plaintiffs were not entitled to recover general damages for mental distress unless such an award was due to physical discomfort or inconvenience but even then, only a "modest" sum is appropriate.

[83] [1991] 1 WLR 1421; [1991] 2 EGLR 152

Chapter Six

Defences to Claims of Surveyors' Negligence

This chapter considers the defences that can be raised by a surveyor facing a claim based on his alleged professional negligence? It should be noted that in negligence, as with many other areas of law, the defences that may be raised are not always absolute defences.

Absolute defences and Partial defences

Absolute defences have the effect of completely discharging the defendant's potential liability. In contrast, partial defences merely reduce the extent of the defendant's liability. In practice, where a partial defence is successfully raised, the court will order a reduction in the damages to be awarded to the claimant.

Defence - "Defendant Not Negligent"

The most obvious and effective defence to any professional negligence claim is to establish that the defendant was not negligent in the first place. However, the prospects of evading liability for negligence will depend on whether the evidence adduced shows that the defendant has fallen below the standard of care required of him, thus breaching his duty to take care.

As with all litigation, there is a constant risk that the case may not run to plan and that a trial judge will accordingly find against you. Thus, an awareness of the defences available, statutory or otherwise, is of key importance and must be considered in detail.

Pleading contributory negligence - The Law Reform (Contributory Negligence) Act 1945

The provisions of the **Law Reform (Contributory Negligence) Act 1945** introduced a partial defence which can be pleaded in all claims for negligence in which the defendant contends that the claimant contributed to his own alleged loss. The contributory negligence defence has the effect of apportioning the liability for the loss suffered between the defendant and the claimant jointly. Section 1(1) of the 1945 Act reads:

> *'Where any person suffers damage as the result of his own fault and partly of the fault of any other person or*

persons, a claim in respect of that damage shall not be defeated by reason of the fault of the person suffering the damage, but the damages recoverable in respect thereof shall be reduced to such extent as the court thinks just and equitable having regard to the claimant's share in the responsibility of the damage.'

In the context of professional negligence actions, contributory negligence seems to be raised as a defence in two types of situation. First, the defence may be used where the defendant believes that the claimant has acted unreasonably by relying on the particular advice in question.

Secondly, there are situations where the claimant's reliance on the advice is not contested but the defendant will seek to establish that unreasonable conduct on the part of the claimant or other external factors influenced the claimant's decision to enter into the transaction.

In any event, those surveyors' negligence cases to date, in which contributory negligence has been pleaded, are predominantly cases where the claimant is a commercial lender. Contributory negligence is undoubtedly easier to prove against a lender as such claimants will be deemed to be more commercially minded than an individual purchaser and will have had the benefit of experience in dealing with property transactions.

Banque Bruxelles Lambert SA v. Eagle Star Insurance Co Ltd [1994]

The case of **Banque Bruxelles Lambert SA v. Eagle Star Insurance Co Ltd [1994]**[84] was effectively the first case in which a negligent valuation surveyor succeeded in obtaining a reduction in the damages payable, on the ground that the plaintiff was guilty of contributory negligence. In Banque Bruxelles Lambert the defendant surveyor had valued the property in question at a much higher price than the amount, which the proposed borrower had provisionally purchased the property for.

Generally, it would seem absurd for a negligent surveyor to argue that his client has acted unreasonably by relying on the information contained in the

---

84 [1994] 2 EGLR 108

survey report but in **Banque Bruxelles Lambert** that exact argument succeeded. The court held that the lender, having been aware of the discrepancy, had not acted prudently by going ahead with the loan. As a result, the damages payable to the plaintiff were reduced by 30 per cent, a figure which reflected their apportionment of the liability.

Nyckeln Finance Co Ltd v. Stumpbrook Continuation Ltd [1994]

Similar facts were seen only a matter of months later by the same court in the case of **Nyckeln Finance Co Ltd v. Stumpbrook Continuation Ltd [1994]**[85] in which a 20 per cent reduction of damages was made for contributory negligence.

Failure to use reasonable care - South Australia Asset Management Corporation v. York Montague Ltd [1997]

More recently in the case of **South Australia Asset Management Corporation v.**

[85] [1994] 33 EG 93

York Montague Ltd [1997][86] the High Court held at first instance[87] that "contributory negligence does not involve a breach of duty owed by a plaintiff to a defendant, but the failure by a plaintiff to use reasonable care to protect its own interests." In the present case the court held that it was just and equitable to make a 25 per cent reduction in damages because the plaintiffs' had:-

(a) failed to make a sufficiently rigorous examination of the valuer's report;

(b) failed to reassess the risk classification of the loan;

(c) failed to give explicit instructions to the defendants.

Claimant's duty to mitigate loss - First National Bank Plc v. Andrew S Taylor (Commercial) Ltd [1997]

In another more recent High Court decision, **First National Bank Plc v. Andrew S Taylor (Commercial) Ltd [1997]**[88], contributory negligence was

[86] [1997] AC 191; [1996] 2 EGLR 93
[87] [1995] EGCS 71
[88] [1997] PNLR 37

raised by the defendant on the basis of the plaintiff's alleged failure to investigate the borrower and guarantor and in not taking proper care to protect the loan.

In any event, the facts in First National Bank were such that the defendant surveyor was held not to be negligent. The trial judge did however state, in *obiter*, that the plaintiff had failed to mitigate its loss and therefore had it been necessary, the court would have reduced the damages payable from £130,000 to £55,000, reflecting a 75 per cent apportionment of liability on the part of the plaintiff.

As already mentioned, a defendant surveyor who asserts contributory negligence in a claim by a purchaser (rather than a lender) may encounter problems in establishing that the lay purchaser was partly negligent in relying on the surveyor's advice.

Having said this, a contributory negligence defence raised on the basis of the claimant's failure to mitigate loss may, in certain circumstances, be just as cogent an argument against a purchaser or borrower as it would be against a commercial lender.

Contributory negligence - Platform Home Loans Ltd v. Oyston Shipways Ltd [1998]

The most recently reported surveyors' negligence case which involved the issue of contributory negligence is **Platform Home Loans Ltd v. Oyston Shipways Ltd [1998]**[89]. The defendant's main argument in that case was that the plaintiff lender had followed an imprudent lending policy.

At first instance the trial judge found that the plaintiff had not taken into account the borrower's ability to service the loan and had not pursued the borrower when details relating to the original price of the property were omitted from the loan application. Accordingly, the defendant's liability was reduced by 20 per cent to reflect the contributory negligence of the plaintiff.

On appeal, the Court of Appeal upheld the trial judge's finding of contributory negligence. Where the decision to loan monies had been reached in

[89] [1998] Ch. 466; [1998] 4 All ER 252

consequence of a lending policy sufficiently imprudent to amount to fault the question of reducing the damages for which the valuers were liable could arise.

The effect of the Limitation Act 1980

It should be evident by now that the liability of a negligent surveyor is limited to those parties who have a contractual relationship with the surveyor or third parties who have suffered loss as a result of that surveyor's negligence. However, this common law restriction on liability is not the only restriction in place.

The Limitation Act 1980 (as amended by The Latent Damage Act 1986)

The provisions of the **Limitation Act 1980** (as amended by The Latent Damage Act 1986) stipulate the maximum time limits within which claimants must commence proceedings. Failure by the claimant to issue a claim form prior to the expiration of the appropriate time period results in the action being statute-barred.

Although this statutory provision is not strictly a

defence, in cases where the sequence of events or timing of events is disputed the defendant will often raise the limitation argument in an attempt to evade liability, whether it be for breach of contract or breach of a common law duty to take care.

Sections 2 and 5 The Limitation Act 1980

Section 2 of the **Limitation Act 1980** provides that 'an action founded on tort shall not be brought after the expiration of six years from the date on which the cause of action accrued'. Similarly, section 5 of the same Act lays down that the time limit for actions founded on contract is also six years from the date on which the cause of action accrued.

Commencement of limitation period - Byrne v. Pain & Foster [1999]

The question of when time begins to run for the purposes of calculating the limitation period was recently considered by the Court of Appeal in **Byrne v. Pain & Foster [1999]**[90]. It was held that a cause of action accrued when the plaintiff first suffered actual damage and in the present case the plaintiffs

[90] [1999] 1 WLR 1849; [1999] 2 All ER 400

had become irrevocably committed to the purchase at the date of exchange of contracts rather than completion. Therefore, despite having issued proceedings within six years of completion but after six years of exchange of contracts, the plaintiffs' action was time-barred.

Section 14A(4)(b) The Limitation Act 1980 - An alternative three year latent time limit

However, a common problem with negligence cases is that the negligence, or the damage caused by the negligence, only becomes apparent some considerable time after it has actually 'accrued'. Thus, section 14A(4)(b) provides that a time limit of three years can operate if it will expire later than the standard six-year limit[91].

Subsection (5) of the same section qualifies the requirements for the special time limit to be valid. It reads :-

[91] However, another provision of The Latent Damage Act 1986 has been inserted into The Limitation Act 1980 which states that a maximum 15 year limit will override the special limit under s14A(4)(b)

'...the starting date for reckoning the period of limitation under subsection (4)(b) above, is the earliest date on which the plaintiff or any person in whom the cause of action was vested before him first had both the knowledge required for bringing an action for damages in respect of the relevant damage and a right to bring such an action.'

Subsection (6) then continues by explaining that "the knowledge required for bringing an action" means both knowledge of 'the material facts about the damage' and other facts such as the defendant's fault in causing the damage or indeed, the identity of the defendant.[92]

Finance for Mortgages Ltd v. Farley & Co [1996]

The case of **Finance for Mortgages Ltd v. Farley & Co [1996]**[93] provides an example of the limitation defence being used successfully by a surveyor accused of negligence. The plaintiffs had purported to extend the validity of the writ and it was finally served on the defendants on 20th December

[92] see Jackson & Powell on Professional Negligence pp.70-71 for a more detailed analysis

[93] [1996] EGCS 35

1993 which was outside the primary six year limit which ran from the date on which the mortgage transaction was completed. The court held that there were no proper grounds to extend that particular writ and so the plaintiffs sought to issue a new writ on 20th April 1994.

The plaintiffs could only proceed on the second writ if they were entitled to rely on the special time limit under section 14A(4)(b). Unfortunately for the plaintiffs, the evidence showed that their knowledge of the valuation's inaccuracy should have been apparent by mid-March 1991. Therefore, the Queen's Bench Division held that the three-year special time limit had, in fact, expired by 31st March 1994 at the very latest and thus the plaintiffs' action was statute-barred.

The Mortgage Corporation v. Lambert & Co [1999]

In the very recent case of **The Mortgage Corporation v. Lambert & Co [1999]**[94], the claimant lender was allowed to proceed with an action commenced some 6½ years after the date of the valuation in

[94] [1999] 3 EGLR 59

question, by virtue of section 14A. The evidence showed that the claimant had been in receipt of estimated values of the property prior to three years before the issue of the claim. However, it was held by the Chancery Division judge that because those estimates were given by unqualified persons for purposes other than valuation, section 14A could be relied upon and thus the claim was not statute-barred.

Chapter Seven

The Perspective of the Practising Surveyor

The rulings in **Watts v. Morrow**[95] and **South Australia Asset Management Corporation**[96] have somewhat reduced the sting in the tail for surveyors facing claims for professional negligence, by limiting the damages recoverable against them. However, valuation and structural surveys are still considered by many to be risky business for the general practice surveyor. With an increasing number of negligence actions coming to court in recent years the surveyor must now, more than ever, be aware of the legal implications of the work which he is undertaking.

There are various ways in which a practising surveyor can, to a certain extent, safeguard himself against liability for professional negligence. Some of these precautions are legal in nature whilst others are merely practical measures approved by the Royal Institution of Chartered Surveyors (or other

[95] [1991] 1 WLR 1421; [1991] 2 EGLR 152
[96] [1997] AC 191; [1996] 2 EGLR 93

associations) to help avoid negligent survey reporting.

The reasonable use of disclaimers

A disclaimer is generally used when a party is seeking to exclude themselves from liability for a breach of a duty of care in the tort of negligence. Disclaimers have a wide range of applications in the law but in the area of surveyors' negligence the main function of a disclaimer is that it purports to give notice that the surveyor will not incur any liability.

Report "without responsibility" - Hedley Byrne & Co v. Heller & Partners [1964]

In the pioneering case on negligent misstatement, **Hedley Byrne & Co v. Heller & Partners [1964]**[97], the presence of a disclaimer, which said that the respondents had given the report "without responsibility", prevented them from being held liable for the loss.

Two more recent House of Lords' cases, which actually involved negligent surveyors, considered the use of disclaimers but in greater detail than in **Hedley Byrne**. The

[97] [1964] AC 465

cases of **Smith v. Eric S Bush** and **Harris v. Wyre Forest District Council [1989]**[98], which were heard together, both involved exclusion clauses which purported to disclaim tortious liability.

Unfair Contract Terms Act 1977 - Smith v. Eric S Bush and Harris v. Wyre Forest District Council [1989]

In **Harris** the plaintiff, a form which contained a declaration stating that the defendant accepted no responsibility for the valuation survey, which had been carried out to assist themselves and not the plaintiff. In **Smith** there were similar facts in that the plaintiff had also signed an agreement containing a disclaimer.

The House of Lords, in coming to a decision, had to consider whether each of these disclaimers came within the provisions of the **Unfair Contract Terms Act 1977**. Section 2(2) of the Act states that :-

> *'...a person cannot so exclude or restrict his liability for negligence except in so far as*

[98] [1989] 2 All ER 514

the term or notice satisfies the requirement of reasonableness.'

Surveyor has burden of proving reasonableness

In reference to this section Lord Griffiths explained that the burden is on the surveyor to establish that it is fair and reasonable that he should be allowed to rely on his disclaimer. In any event, the House of Lords suggested that the four criteria, as listed below[99], should be considered in deciding whether a particular disclaimer satisfies the test of reasonableness within the meaning of the **Unfair Contract Terms Act**:

Test of reasonableness

- Were the parties of equal bargaining power?
- Would it have been reasonably practicable to obtain the advice given from an alternative source, taking into account costs and time?
- The difficulty of the task undertaken.

[99] summarised list taken from 'Professional Negligence & Insurance Law' p. 27

- The practical consequences of the decision, taking into account the fees charged for the work, the ability to bear the loss and any insurance cover.

In **Smith** and **Harris** both Lord Templeman and Lord Griffiths gave their views as to the validity of the disclaimers in question. In his speech Lord Templeman reflected on the poor state of the property market at that time. He held that, in the absence of a statutory provision to the contrary, valuation surveyors were not entitled to rely on general disclaimers which purport to exclude the common law duty of care owed by valuation surveyors to the purchasers of real property.

Lord Griffiths however, was more reluctant to rule out the use of disclaimers altogether. Although he found that the clauses used in the present cases were unreasonable, he was prepared to entertain the proposition that in the case of expensive or possibly industrial properties "prudence would seem to demand that the purchaser obtain his own structural survey". His Lordship continued by suggesting that in these circumstances, it may be reasonable for valuation surveyors to disclaim or limit their liability.

Conditions of engagement

It has been suggested that the surveying profession as a whole generally accepts that a duty not to be negligent is an inherent part of the professional - client relationship. On this basis, it may be the case that disclaimers are, in actual fact, rarely found in survey reports.

Instead of using disclaimers which the **Unfair Contract Terms Act** may render ineffective in any event, every prudent surveyor should insert terms of engagement which 'define clearly the limits of the service to be provided.'

Such terms are commonly known in the surveying profession as the 'conditions of engagement'. Many firms have now adopted the practice whereby a copy of the conditions of engagement are sent to the client prior to the survey being carried out. Some firms even insist that the client signs a declaration to the effect that he is aware of the conditions of engagement and intends to be legally bound by them. Other firms merely send the conditions out on the basis that if the client has a problem with them, or any part of them, he should say so prior to the work being undertaken.

A typical example of a practising surveyor's conditions of engagement can be seen in Appendix 1[100]. In effect, the conditions of engagement seek to inform the client what will be inspected during the survey and perhaps more importantly, what will not be inspected. This measure will, in theory, protect the surveyor from dissatisfied clients coming back, months or even years later, complaining that he has failed to do his job properly, which is all too often the first indication that legal proceedings will follow.

In addition to an outline of the items covered by the survey, the conditions of engagement often includes a list of the exclusions. The more common of such exclusions are as follows :-

---

[100] Appendix 1 contains the 'Terms of Engagement' of Vegoda & Co, Chartered Surveyors.

- The survey will not purport to examine parts of the structure of the property which are covered, unexposed or inaccessible.

- Floor coverings are not normally raised and furniture is not normally moved.

- Services, including electricity, gas and water, will not usually be tested although an overall impression of the state of the fittings will be given.

- Any measurements taken are included in the report for descriptive purposes only and therefore their accuracy is not guaranteed.

- The surveyor will usually reserve the right to call for specialist assistance where it is considered necessary (e.g. a structural engineer's report)

- Estimates given in the schedule of repairs are intended to give a rough guide to the costs involved and thus contractor's estimates should also be sought.

- Finally, the report is a confidential document which is made solely for that particular client. Therefore the survey report may not be sold or lent to third parties to whom the surveyor will not be held liable.

The RICS guidance booklet gives some valuable information and advice on reporting and furthermore, it specifically gives details of recommended conditions of engagement. However, it should be noted that the recommended wording for use in the conditions is tailored towards the degree of inspection required by a Home Buyers' Survey and Valuation report. Therefore, a surveyor who undertakes a full structural survey may find that the conditions of engagement detailed in the booklet are somewhat inadequate in places.

The application of British Standard 5750

The standard of care required of the surveyor is that of the reasonable man who professes to be skilled in the area of surveying. It is therefore not surprising that in many cases involving surveyors' negligence the professionalism of the surveyor is brought into question. Indeed, the plaintiff may put forward evidence to show that the surveyor could have used more effective procedures than those which he used to carry out his work. 'Quality assurance' is a term which has become common in all types of professions and industries during the recent past.

The BSI's introduction of British Standard 5750 in early 1992 was aimed at implementing a quality assurance standard across a wide spectrum of British businesses and professional bodies. In an article in the May 1994 issue of Professional Manager[101], the director of the BSI explained that "The benefits of applying BS5750 are real; it will save you money - because your procedures will be more soundly based and efficient; it will ensure satisfied customers - because you have built-in quality at every stage; and it will reduce the time consuming re-working of procedures."

In relation to the surveying profession, firms which have been registered as BS5750 approved can generally expect to win more business as well as experience improved

[101] see 'Debating the case for quality' by John Seddon

administrative efficiency. Indeed, in a survey highlighted by the article in Professional Manager, 69 per cent of the businesses consulted claimed to have improved procedural efficiency. It should be borne in mind though, that implementing BS5750 will not prevent a surveyor from failing to detect a serious defect in the property which he is surveying.

Recognition by the courts - Bere and Parchment v Slades [1990]

In cases such as **Bere and Parchment v. Slades [1990]**[102], the courts have tended to look favourably on the fact that the surveyor in question has conducted his services in a professional manner.

Although BS5750 (and the similar ISO9000) only offers the practising surveyor the benefit of having thorough work procedures, this measure may, in limited circumstances, safeguard the surveyor from liability for professional negligence.

It would seem though, that the best way in which a surveyor can safeguard himself against negligence claims, is to bear in

---

[102] [1990] 6 Const. LJ 30

mind the test used by the courts for assessing the advice given by professional defendants, each and every time he carries out a survey.

Chapter Eight

Insuring Against the Risk

It has been considered how a practising surveyor might attempt to reduce the extent of his or her liability, both by way of legal and practical means. However, it would be naive to think that these measures could rule out the possibility of liability for negligence completely. To coin an old phrase, 'nobody is perfect', and it is inevitable that there will be instances where normally competent surveyors fall foul of the standard of care required of them, either by arriving at an unrealistic valuation or by failing to detect a substantial defect in a property.

It is for this reason that the RICS insist upon all of their members having a minimum level of insurance. This type of specialist insurance is known as professional indemnity insurance.

Insurance requirements of the RICS

The Royal Institution of Chartered Surveyors have had a rule for many years which obliges each of its practising members to take out professional indemnity insurance. It would appear that this is primarily to safeguard the reputation of the profession but at the same time it has the effect of protecting its members from otherwise financially crippling damages payments and legal costs. The minimum level of cover required by the RICS is actually very wide, especially when one considers the policies adopted by other professional bodies.

The RICS insist that a basic policy must include cover for claims arising from any civil liability. Traditionally, the cover on a professional indemnity policy would have only been for tort based liability such as a negligent breach of professional duty but the level of cover required by the RICS includes claims arising from contractual and other civil liability as well.

At present a practising surveyor is able to buy a policy from whoever he or she chooses but the RICS have plans to formulate a mandatory insurer list which would give details of approved brokers. The need for such a scheme lies with the fact that in the past, some firms of surveyors have been deemed to be uninsurable. This posed the problem of whether the RICS should expel these members for failing to comply with the indemnity rule.

Another planned innovation of the RICS, although not a new idea, is the creation of an 'assigned risks pool', whereby an amalgamation is formed of all those underwriters wishing to insure high risk surveyors. Each of the assignees is then allocated a share of the risk depending on the proportion of the premiums they receive.

The high price of insurance cover

Surveyors who undertake work such as structural surveys and valuation surveys have long been regarded by underwriters as high risk customers. Therefore, both the policy premiums and the mandatory excesses payable on each claim are often high. The scope of information required from a surveyor wishing to take out a professional indemnity policy is usually quite extensive.

There were a string of cases regarding negligent valuations that came to court following the fall of the property market in the early nineties[103]. Not surprisingly, premiums for professional indemnity policies rose dramatically in order to accommodate the increased number of claims against negligent surveyors.

This sudden trend in suing valuation surveyors meant that firms with a record of claims against them found it difficult to get cover at all. Many insurance brokerages attempted to create a market for these firms but as one broker explained, *"it's a bit like selling ice creams in a heat wave, you don't have to do it cheaply."*

'attritional' losses

The main reason that premiums and excesses stay high is that many of the claims made against surveyors are written off as 'attritional' losses. An illustration of this concept is as follows :-

> *A newly married couple buy a house. This is the single biggest investment they have ever made. Upon moving in*

---

[103] see United Bank of Kuwait v. Prudential Property Services Ltd [1994] 30 EG 103,
Banque Bruxelles Lambert SA v. Eagle Star Insurance Ltd [1994] 2 EGLR 108 and
Nykeln Finance Co Ltd v. Stumpbrook Continuation Ltd [1994] 33 EG 93

they discover some relatively minor defects undetected by the surveyor. Funded by legal aid they sue the surveyor for their £5,000 loss, but the legal costs, which would be incurred in fighting the case, cannot be justified by the underwriters. Even if the insured won, he would not be awarded costs against a legally aided plaintiff. Therefore the claimants will often be paid off in an 'out of court' settlement even if the surveyor may not have been negligent in failing to spot the defect.

'Claims made' basis

Another cause of the high cost of professional indemnity insurance for surveyors is the fact that standard policies operate on a 'claims made' basis. This means that from the date that the policy starts to the date that it ends the insured will be covered for any claim made against him or her, regardless of whether the cause of the action occurred before the policy commenced.

By the same token, if a surveyor negligently values a property during the period of cover but the subsequent claim is not made until after that policy expires, the surveyor will not be covered by the expired policy. This is the reason why proposal forms ask for details of previous work undertaken by the surveyor.[104]

The concept of instructing a surveyor as 'buying insurance'

An insurance broker with whom I spoke thought that the majority of claims brought against surveyors are justified in so far as the surveyor has not been professional in his or her work. He believes that the root of the problem is that many mortgage lenders are in such a rush to process each loan that they put extreme pressure on valuation surveyors to arrive at a value as quickly as possible.

Many surveyors cannot afford to turn down the work (especially from regular clients such as banks and building societies) even if they don't have the time to devote to it and thus the standard of service that they give is often less than adequate. This type of scenario was particularly common in the property 'boom' of the mid to late eighties and it was not until many of these 'second gear

---

[104] see the proposal form in Appendix H for examples of this

valuation' cases came to court that the surveyors responsible were caught out.

A phenomenon which appears to have emerged, in light of the increasing amount of litigation involving negligent surveyors, is the idea that people who instruct either structural or valuation surveyors are, in fact, merely buying insurance. As is mentioned above, every surveyor must have professional indemnity insurance and therefore if he or she fails to do the job properly any loss incurred by the client will be covered by the surveyor's insurance policy.

The House of Lords showed an understanding of the economic realities of the surveying profession in the joint cases of **Smith v. Eric S Bush** and **Harris v. Wyre Forest District Council [1989]**[105]. The House held that one of the four criteria which should be used to assess a disclaimer's reasonableness was the ability of the defendant surveyor to bear the loss by means of insurance cover. This endorsement of professional indemnity insurance by the House of Lords lends support to

[105] [1989] 2 All ER 514

the notion that many mortgage lenders, and homebuyers for that matter, regard the instruction of a surveyor as an insurance against the *caveat emptor* risk.

Chapter Nine

Conclusions

Does a doctrine of professional negligence exist?

The growth of professional negligence[106] litigation strongly suggests that the law of liability for negligent misstatements has evolved in line with general principles of tort and, if this is the case, it is likely that the **Hedley Byrne** principle will continue to underpin the body of law in future.

Indeed, the evidence is that the **Hedley Byrne** authority is finally establishing itself as the central principle of professional negligence law irrespective of whether the negligence in issue takes the form of misstatement or misfeasance.

Spring v. Guardian Assurance plc [1994] and Henderson v. Merrett Syndicates Ltd [1994]

In response to the question of whether a doctrine or general theory of professional negligence exists, it would

---

[106] see 'Is there a general theory of professional negligence liability in tort?' by K M Stanton Professional Negligence, Volume 10, Number 4, 1994

appear **Smith** and **Harris** has paved the way for more recent developments, such as the House of Lords' rulings in **Spring v. Guardian Assurance plc [1994]**[107] **and Henderson v. Merrett Syndicates Ltd [1994]**[108], which point to the presence of general theories of negligence liability.

In both of these cases Lord Goff stated, with reference to the principle laid down in **Hedley Byrne**, that when dealing with a professional person there is an assumption of responsibility and thus it follows that they will owe an implied duty of care.

The surveyor's standard of care extended?

At a lecture I attended in 1996, entitled 'Structural Surveys - Residential Property', one of the delegates raised an interesting issue relating to the standard of care required of a surveyor. He asked whether a surveyor could be found negligent for failing to warn the prospective purchasers of plans to construct a new road link in the vicinity of the property being inspected, a factor which could well affect the property in terms of

---

[107] [1995] 2 AC 296: [1994] 3 All ER 129
[108] [1995] 2 AC 145; [1994] 3 All ER 506

economic value and possibly even structural stability.

Busy flight path - Farley v. Skinner [2000]

Another delegate had also heard of an action brought against a surveyor on the grounds that he had failed to warn the prospective purchasers of the busy flight path that passed over the property being surveyed. This was considered in the very recent case of **Farley v. Skinner [2000]**[109], in which the claimant purchaser specifically instructed the defendant surveyor to advise whether the property might be affected by aircraft noise.

At first instance, Judge Peter Baker QC held that the defendant surveyor had been negligent for failing to alert the claimant to the presence of a nearby navigation beacon used by air traffic control at Gatwick Airport and the associated aircraft noise that would be likely in that area.

The judge accordingly made an award of damages and interest to compensate the claimant for the impairment of his use and enjoyment of the property. The Court of Appeal subsequently

---

[109] [2000] EGCS 52

held, by a majority of two to one, that damages were only recoverable, in this type of case, for physical inconvenience and discomfort caused by the breach and mental suffering directly related to that inconvenience and comfort.

Contract for pleasure, relaxation or peace of mind - Jarvis v. Swan Tours Ltd [1973]

Furthermore, Lord Justice Stuart-Smith explained that damages for loss of use and enjoyment could only be recovered in an action based on breach of contract where the contract was primarily entered into for the purpose of pleasure, relaxation or peace of mind. The case of **Jarvis v. Swan Tours Ltd [1973]**[110] was given by the court as an example of the type of contract under which damages for loss of use and enjoyment would be recoverable. The defendant surveyor's appeal was allowed and the claimant was ordered to repay the £10,750 awarded to him in damages.

Interestingly, it was only the first instance judge's award of damages which was appealed to

[110] [1973] QB 233

the Court of Appeal and not the underlying finding of negligence itself. However, this may be due to the fact that the claimant expressly instructed the defendant surveyor to identify any problems of aircraft noise given the property's proximity to Gatwick airport.

It still remains to be seen whether the surveyor's standard of care should be limited to the fabric of the property where there has been no specific instruction to check a peripheral factor. As a matter of prudence, such ancillary matters should be included in the solicitor's enquiries before purchase.

After all, if a proposed new road link or the flight path are partly the reason why the vendor is selling, then failure to disclose this information in the solicitor's enquiries may well constitute a negligent or even fraudulent misrepresentation[111], for which the subsequent purchaser will have legal recourse.

The scope for Law Reform

In recent years the Department of Trade & Industry have on occasions instructed the Law Commission to carry out

[111] see The Misrepresentation Act 1967, section 2(1)

feasibility studies with a view to reforming the law in relation to the liability of professionals generally. These moves come in response to growing concerns surrounding the increase in negligence claims against professionals and the subsequent inflated costs of professional indemnity insurance.

It is widely felt that the increasing tendency towards litigation against surveyors was vastly intensified by the last economic recession in the UK, an inevitable feature of which was a collapse of the then thriving property market.

With property prices once again rising sharply in the UK (particularly in the south-east) during the last 18 months, a slump in the property market is a real possibility at some point in the not too distant future. If this does in fact happen a resurgence of valuation based negligence claims would almost certainly ensue.

It appears to me that the traditionally perceived image of the law as 'scales of justice' is remarkably symbolic of the changes experienced in the development of the law relating to surveyors' negligence.

The common law has done its best to achieve a balance of interests between the protection of the defendant surveyor on the one hand and the interests of the homebuyer on the other. However, there seems little doubt that the work of the general practice surveyor has become an increasingly risky business and thus it may well be time for the legislature to offer the troubled surveyor some timely respite.

Appendix 1

Typical Terms of Engagement

Terms of Engagement

STRUCTURAL SURVEYS - ITEMS NORMALLY COVERED

Damp Proof Course:

We test accessible parts throughout with a moisture meter and report on any dampness found, its degree and action we recommend to be taken.

Decoration:

Description:

Drainage:

Where possible drain covers will be lifted and the drains examined at chambers. We carry out a visual inspection. We cannot always raise bolted down covers; covers rusted to frames; broken covers; tiled covers or covers where there is risk of damage to the finish in raising them.

Eaves:

A visual inspection is carried out.

Electricity:

Power points are normally tested, where accessible with a Martindale Ring Main Tester and wiring is exposed at randomly chosen light and power points.

Exclusions:

All survey reports contain the following exclusion clauses:-

We have not examined woodwork or other parts of the structure which were covered, unexposed or inaccessible and we are therefore unable to report that the building is free from defect.

We would add that we were not able to carry out investigations to determine whether high alumina cement, calcium chloride or any other deleterious materials were used during the construction of the premises and we are therefore unable to report that the building is free from risk in this respect.

This report is only for the use of the person to whom it is addressed and no liability can be accepted to third parties. It may not be sold, lent or copied without express permission in writing and may not be relied upon until payment of our fee account in full.

Measurements are included for descriptive purposes only and their accuracy is not therefore guaranteed.

VAT is currently chargeable on most building work and estimates given are exclusive of VAT and are rough and ready guides only. We suggest that contractor's estimates are sought before entering into any commitment.

Where services such as gas; water; and electricity have been disconnected, there is no guarantee that the suppliers will reconnect without requiring major work.

Inquiries should be made before entering into a binding commitment.

External Timber:

Comments on the soundness or other wise of timbers examined and remedial action, if any, recommended.

Fences and Gates:

Details of these and their condition.

Floor Structure and Air Bricks:

Adequacy of ventilation and freedom from checkable structural inadequacies. Floor covering are not normally raised and furniture is not normally moved. Should you wish we can arrange for builders to attend and raise and relay floors for inspection subject to an additional charge and an indemnity from you against damage.

Flues:

Not usually inspected.

Garage and Outbuildings:

Gulleys:

Open gulleys are inspected where conditions permit. Recommendations of necessary repairs will given or alternately any works required to establish their condition. We do not normally unscrew bolted down gulleys.

Heating:

The boiler is normally examined and fired where it is possible to do so. Covers will be removed where they can be without removing knobs or dismantling part(s) of the boiler. Casing will not normally be removed. The nominal output will be established if possible and we will normally venture an opinion if we think it very under powered. We do not always test every single radiator.

Internal Woodwork:

Photographs:

Usually included in a survey report but not in abbreviated Homebuyers type reports.

Plan:

A line drawing is usually included for identification purposes but not in Homebuyers type reports.

Plaster:

We describe the condition of plaster, and necessary repairs so far as we can establish it.

Plumbing:

Examined where accessible and defects found are reported and where appropriate, remedial action recommended.

Rainwater and Soil Goods:

Roof:

A description of the roof or roofs, their condition and whether they are approaching the end of their useful life or need renewal. If safely accessible with our 12'0" ladder we would normally inspect the interior of the roof and the exterior using optical instruments where appropriate.

Services:

Whether there is gas, water, electricity and main drainage and where the services are.

Settlement:

An indication of movement in the property or on the basis of the facts available at the time of our single inspection whether we think movement is more than averagely likely to occur.

Specialists:

We reserve the right to call for specialist reports where we consider it appropriate.

Trees:

Any trees or plants situated within the site or elsewhere which might affect the property or its neighbours.

Valuation:

Not normally included in the survey report.

Value for Insurance Purposes:

Not normally included in a survey report.

Walls:

Windows and Doors:

BUILDING AND PLANNING WORK

This may (but does not necessarily) include:-

Surveying and measuring the existing structure and noting details required for Applications for Statutory Consents such as positions and depths of drains; positions of services and trees etc. Drawing up plans of the existing structure. Drafting proposals in accordance with your instructions and if necessary revising the plans following discussions. Submitting applications for Planning Consent, subject to payment of the statutory fee by you and our fees and disbursements accrued to date. Disbursements would include amongst other things, such items as dye line prints and Ordnance Survey Maps required by Local Authorities with every application.

If the planning process is successfully completed or if you require us to do so concurrently we can apply for consent under Building Regulations. Fees are payable to the local Planning authority on making a planning application, and fees are payable to the Building Control Officers on submission of plans and again following their first site inspection.

We cannot guarantee that an application for Planning Consent will be granted. Applications for Building Regulations Consent will normally be approved provided the proposal complies with the regulations. Sometimes applications are returned for amendment and in view of the current delays in obtaining consent, a building notice is usually served following submission of plans and works may commence three days after service of the notice.

It is sometimes necessary to instruct professionals in other disciplines and we will advise if we think this is necessary. For example, an engineer's calculations are often needed before applying for consent under Building Regulations. Such professional advisers will be instructed by you, paid by you and be responsible to you for their work. We do not take responsibility for work carried out by such advisers.

Should you require us to:

- draft a specification
- go out to tender,
- let the contract,
- Administer the contract
- value stage and final payments.

- then we charge 12½% of the contract sum plus disbursements and VAT. Credit will be given for interim payments of fees.

Interim charges at stages are based upon the time spent, plus disbursements, plus VAT and invoices will be submitted and become due from time to time. Where %age charges are disproportionately small in relation to the work done, charges will be based on the time expended. Minimum fee £500 plus disbursements and VAT.

RELEASE OF PLANS AND REPORTS

Payment is required BEFORE release. Should work be released without prior payments for any reason the terms of payment set out in this document will apply.

TIME CHARGES

Principal's time is charged at a £(instruct me and find out) per hour, assistant's at £30 per hour, plus disbursements plus VAT. Abortive work is charged on a time and disbursements basis.

DISBURSEMENTS

Disbursements will be charged at cost. Charges will also be made for sending or receiving telephone calls and facsimiles.

VAT

Payable at the standard rate for the time in being.

VALUATIONS

0.2% of the valuation with a minimum fee of £500 plus VAT. Or such other minimum fee as is specified when accepting the instruction.

RENT REVIEWS

10% of the average rent agreed. Minimum fee £500 plus VAT. For arbitration an additional fee based upon time taken for preparing the submission and for attendances and any other work required to finalise the fixing of the rent.

LITIGATION - EXPERT WITNESS

Charges according to time spent plus disbursements. See time charges above. Court work at 150% of time charges.

SCHEDULES OF DILAPIDATIONS

Charged according to time taken plus disbursements and VAT. Contract Administration is not included in this work. See 'Building and Planning Work' above.

CAPITAL ALLOWANCE CLAIMS

1% of the first £½ million
0.75% of the next £½ million
0.5% of the residue
Minimum fee £1,250

PLANNING SUPERVISOR

1% of the contract sum. Minimum fee £1,000

TERMS OF BUSINESS

Settlement required within 14 days of the invoice date. Interest will be charged on balances outstanding after 14 days at the compound rate of 3% (variable) per month or part thereof which amounts to almost 38½% APR. Letters & telephone calls chasing overdue accounts £20 each. Recovery costs and our time spent in recovering fees are charged on an indemnity basis. Ongoing instructions will be invoiced periodically. The Contracts (Rights of Third Parties) Act 1999 does not apply.

CASE SUMMARIES

Reports are reproduced by the kind permission of:-

The Incorporated Council of Law Reporting for England and Wales

and

The Estates Gazette.

All rights reserved.

Anns v. Merton London Borough Council
[1978] AC 728

Local authority's duty of can in carrying out their functions in relation to the control of the construction of buildings under the Public Health Act 1936 and building by-laws - Duty to consider whether to inspect, but not absolute duty to inspect in all cases - Liability for lack of reasonable care, however, if inspection is in fact carried out – Authority would he liable for negligent inspection of foundation - Period of limitation for action under Limitation Act 1939 - Date when defects first become apparent - Damages recoverable - Authority's appeal dismissed

BNP Mortgages Ltd v Goadsby & Harding Ltd
Official Referee's Business
[1994] 42 EG 150

Negligence - Mortgage valuation - Whether valuation negligent - Whether valuer made valuation fraudulently - Whether defendants liable for fall in property values

Held:

The defendants' valuer had previously inspected the property on June 6 1989 with other adjoining properties and valued the property for a different lender. He did not, contrary to his evidence, revisit the property in February 1990. On the evidence a proper valuation of the property in February 1990 by an ordinary competent valuer was £180,000, but the highest non-negligent valuation was £200,000. Accordingly the valuation of £245,000 was negligent. However, the defendants had not colluded with the borrowers and had not been fraudulent in producing the valuation. The case was a non-transaction one, the plaintiffs relying on the valuation to make the advance.

The risk of a fall in property values was not a risk in respect of which the plaintiffs placed reliance on the defendants' valuation.

Banque Bruxelles Lambert SA v. Eagle Star Insurance Ltd [1994] 2 EGLR 108

United Bank of Kuwait v. Prudential Property Services Ltd [1994] 30 EG 103,

Nykredit Mortgage Bank plc v. Edward Erdman Group Ltd Formerly Edward Erdman (an unlimited company) [1995] 1 EGLR 129

BNP Mortgages Ltd. v Key Surveyors Nationwide Ltd

BNP Mortgages Ltd v. Goadsby & Harding Ltd [1994] 42 EG 150

Mortgage Express Ltd. V Bowerman and Partners (a firm)

Damages - Measure of damages -Surveyor's report - Plaintiffs lending money, on basis of surveyor's negligent valuation - Subsequent general fall in properly market and default by borrowers in repayment of loans - Plaintiff obtaining possession and selling properties at price below valuation -Whether loss attributable to market fall recoverable

Held,

allowing the appeals in the first, fourth, fifth and sixth actions and dismissing the appeals in the second and third actions, that where a mortgage lender would not, but for the negligent valuation, have entered into the transaction with the borrower he could recover the net loss he had sustained as a result of having done so; that a fall in the market was foreseeable, and since, in such a case, the lender would not have entered into the transaction but for the valuer's negligence and could not escape from it unless and until the borrower defaulted, that negligence was the effective cause of his loss, and a fall in the market was not to be treated as a new intervening cause breaking the link between the valuer's negligence and the damage sustained; and, that, accordingly, on the assumed facts, the first defendant in the first action and the plaintiff mortgagees were entitled to recover damages in respect of the loss they had sustained which was attributable to market fall.

Baxter v. F W Gapp & Co Ltd
[1939] 2 KB 271; [1939] 2 All ER 752

Negligence – Damages – Measure – Valuation of Property for mortgage – Excessive valuation – loss of mortgage – Method of computation.

The plaintiff, relying on a valuation of freehold property by the defendants. advanced money on mortgage to the owner of the property. The valuation was excessive, it having been made without the skill and care which the defendants owed to the plaintiff, and the plaintiff suffered loss owing to the default of the mortgagor :-

Held,

by the Court of Appeal, that, in an action by the plaintiff against the defendants for negligence, the plaintiff's damages were not limited to the difference between the amount of the valuation and the true value of the property at the time of the valuation, but that he was entitled to recover the actual loss suffered by him as a result of his lending the money, including the difference between the sum advanced by him and that received by him when, having entered into possession of the property, he sold it, the amount of interest which the mortgagor had failed to pay, the cost of insuring the property and of maintaining it in repair while it was in the plaintiff's possession, legal charges during that period, the expenses of abortive attempts to sell the property, estate agents' commission on the eventual sale of the property and legal charges in connection with the sale.

Decision of Goddard L.J. affirmed.

Bere And Another v Slades
Official Referee's Business
[1989] 46 EG 100

Negligence - Claim by purchasers of house against valuers for alleged negligence in a valuation for mortgage purposes - Defendants instructed by building society - Inspection carried out at the end of 1984 - Question as to whether there had been a lack of care in following up defect observed in the course of inspection - Standard of care required - Smith v Eric S Bush and Roberts v J Hampson & Co cited - Valuer's duty, to "follow the trail of suspicion" - Plaintiffs' action fails

Plaintiffs, one of whom had occupied a flat in the subject house, decided to join together in purchasing the house when it came on the market - In deciding to purchase it, for the sum of £30,000, they, relied on the valuation report made by, the defendants for the purposes of mortgage on the instructions of a building society. - Plaintiffs paid a fee for the valuation - The report drew attention to evidence of past structural movement, which the valuer regarded as neither serious nor progressive, although he recommended an investigation of the drains to eliminate any leak which might cause trouble – He advised that the timber structure of the roof should be looked at and, in view of wood in the cellar of the house, recommended an inspection for beetle infestation

The judgment does not go into details as to the scale of the troubles or the sequence of events which gave rise to the present action, in which a number of experts gave evidence, some of it conflicting - The judge was satisfied that the plaintiffs had relied on the defendants' report in deciding to purchase the house - There was no dispute as to the standard of care which was applicable; this was clearly laid down in Smith v Eric S Bush and Roberts v J Hampson & Co – The valuer in these circumstances owed a duty to use reasonable care and skill so as not to miss defects which were observable by careful visual examination - If there were specific grounds for suspicion, the valuer had a duty to take reasonable steps to follow the trail of suspicion until he had all the information which it was reasonable for him to have before making his valuation - In the present case the main issue was as to whether the valuer had carried out sufficiently the duty to "follow the trail"

It was conceded that the defective wall construction, consisting of sand and aggregate mixture with some cement, was not a defect which was directly discoverable in the course of a valuation survey of the kind which the defendants were instructed to make - The plaintiffs, however, submitted that there were features which should have put the valuer on notice that serious problems existed which required further investigation - It was suggested that the valuer should not have proceeded with the valuation until further investigations had been made - Examples of the features mentioned were the recurrence of external cracks in a flank wall, some internal cracks, the construction and state of cellar walls and the roof space (into which it was alleged that the valuer did not enter) - The judgment dealt in detail with the evidence under each of these headings and preferred the evidence for the defendants

The judge's conclusion was that, taken together in the context of the building as a whole, its age and state of decorative repair, and having regard to similar houses in the same area, the defects mentioned did not give rise to any duty on the valuer to refrain from proceeding with the valuation or to make further recommendations for investigation beyond those which he had already made - It was therefore unnecessary, to assess damages, but the judge would have accepted the view that the house in its defective condition would have realised a market price of about half that paid by the plaintiffs - The valuer's report was "a thoroughly competent report, carefully done and, I would add, masterly succinct-

Byrne and another v Hall Pain & Foster and others
Court Of Appeal
[1999] 12 EG 165
[1999] 2 All ER 400

Negligence – Surveyors – Tort - Limitation – Negligent survey - Whether cause of action accrues at contract date or completion

Held:

The appeal was dismissed. The issue relating to the correct date at which the cause of action accrued was one that could be decided on an application to strike out under RSC Ord 18 r 19. On exchange of contracts the plaintiffs became irrevocably committed to acquiring the lease of the flat. The cause of action accrued when contracts where exchanged. It was on this date that the plaintiff's acted on the report and suffered damage.

Caparo Industries plc v. Dickman
House of Lords
[1990] 2 AC 605; [1990] 1 All ER 568

Negligence - Duty of care to whom? - Auditor-Appointment by company to audit and certify company's accounts - Statutory duty to make report to shareholders - Another company making take - over bid by initial purchase of shares - Claim that subsequent completion of take-over by purchase of total issued shares made in reliance on negligently made audit - Whether auditor owing duty of care to shareholders - Whether duty owed to non - shareholding investor-Companies Act 1985 (c. 6), ss. 236(1)(2), 237(1)

Held,

allowing the appeal and dismissing the cross-appeal, that liability for economic loss due to negligent mis-statement was confined to cases where the statement or advice had been given to a known recipient for a specific purpose of which the maker was aware and upon which the recipient had relied and acted to his detriment; that since the purpose of the statutory requirement for an audit of public companies under the Act of 1985 was the making of a report to enable shareholders to exercise their class rights in general meeting and did not extend to the provision of information to assist shareholders in the making of decisions as to future investment in the company, and since, additionally, there was no reason in policy or principle why auditors should be deemed to have a special relationship with non-shareholders contemplating investment in the company in reliance on the published accounts, even when the affairs of the company were known to be such as to render it susceptible to an attempted take-over, the auditors had not owed any duty of care to the plaintiffs in respect of their purchase of F Plc.'s shares.

County Personnel (Employment Agency) Ltd V Alan R Pulver & Co
Court of Appeal
1 WLR 916; [1986] 2 EGLR 246

Negligence Claim against solicitors - Advice in regard to underlease Unusual rent review clause

Rent of underlease of two rooms on ground floor of a building was to be £3,500 per annum inclusive of rates etc. for the first five years and thereafter to be, for the periods set out in the headlease, an amount equivalent to the initial rent increased by the same percentages as the mesne landlord's rent for the whole building would be increased under the terms of the headlease - This arrangement might have been reasonable if both initial rents had been set at the market level - In fact the initial headlease rent was below the market level and the initial underlease rent far above it - In addition, when the underlease was entered into the solicitor advising the underlessees, plaintiffs in the action, had no information as to the amount of the headlease rent - After five years the initial headlease rent, which was £2,250 only for the whole building, was increased to £5,800 and the initial underlease rent of £3,500 rose to £9,022 (at the open market value this rent would have been £2,600 inclusive) - Plaintiffs could not continue and had to pay £16,000 to be allowed to surrender their underlease, plus a sum of £2,761, representing the increased rent payable since the review date - Plaintiffs claimed these amounts as damages and in addition a sum of £17,000 lost on a prospective sale of the underlease and goodwill

Held,

reversing the decision of the deputy judge, that there were features which should have impinged on the mind of a reasonably careful and competent solicitor, the main ones being the head lessee's unwillingness to divulge the rent being paid for his lease of the whole building and the very unusual nature of the review clause - The clients should have been warned of the risks involved and, indeed, advised not to enter into the underlease on these terms - It was negligent not to give such advice - As regards damages, the "diminution of value rule" was not here appropriate - The true measure was the actual cost of extricating the plaintiffs from the consequences of the negligent advice - There was evidence that (his cost was £18,761 - The additional claim for £17,000 was speculative and should be viewed with caution - The case would be remitted to a Chancery. master for the assessment of damages - Appeal as to liability allowed

Dodd Properties (Kent) Ltd And Another V Canterbury City Council And Others
COURT OF APPEAL
(1979) 253 EG 1335

Assessment of damages Damage caused to garage premises by works carried out In construction of adjoining multi-storey car park - Claims by owners and lessees of garage premises against city council, main building contractors and subcontractors - Main issue date as at which damages should be assessed - Whether date at which cause of action arose or date when repairs could reasonably have begun - If later date correct in principle, should it be taken to be a date in 1970, as Cantley J held, or the date In 1978 when the plaintiffs' action was heard

Held by Court of Appeal

that the later date was correct - It was reasonable to defer repairs until the defendants' liability had been established - Judge wrong in considering that the Lieshosch decision compelled him to select the earlier date - Criticism of Philips v Ward - Appeal allowed - Large difference in amount of damages

Donoghue v. Stevenson [1932] AC 562

Negligence - Liability of Manufacturer to ultimate Consumer - Article of Food - Defect likely to cause Injury to Health.

By Scots and English law alike the manufacturer of an article of food, medicine or the like, sold by him to a distributor in circumstances which prevent the distributor or the ultimate purchaser or consumer from discovering by inspection any defect, is under a legal duty to the ultimate purchaser or consumer to take reasonable care that the article is free from defect likely to cause injury to health:-

So held,

by Lord Atkin, Lord Thankerton and Lord Macmillan; Lord Buckmaster and Lord Tomlin dissenting.

George v. Skivinglon (1869) L. R. 5 Ex. 1 approved.
Dicta of Brett M.R. in Heaven v. Pender (1883) 11 Q. B. D. 503, 509-11 considered.
Mullen v. Barr & Co., Ld., and M'Gowan v. Barr & Co., Ld., 1929 S. C. 461 overruled.

Eley And Another v King & Chasemore
Court of Appcal
[1989] 22EG109

Negligence - Survey, report - Cracks in structure - Condition of roof - Surveyors' advice that purchasers should insure against subsidence - Subsequent discovery of serious defects in roof and complaints of movement in main structure - Court of Appeal uphold judge's decision that surveyors had not been negligent - A comment on pleading

Appellants, husband and wife, purchased a house at Burgess Hill in 1984 after receiving a survey. report from a partner in the respondent firm - The report drew attention to numerous cracks in the wall structure and elsewhere; pointed out that the house stood on shrinkable clay; noted that there was a fir tree in the front garden which might cause trouble; and advised the appellants that insurance protection should be obtained, if possible, although there might be difficulty about it - As regards the roof, the surveyor noted that the felt had been laid in an "unconventional" way, over instead of under the battens, possibly, after bomb damage; that there was no apparent dampness coming from the roof; and that in genera] his impression was that the roof covering was effective - The surveyor had looked inside the roof space but had not been on the roof, not having had long enough ladders to inspect the valley gutters - The appellants purchased the house for £87,500, a little more than the £86,000 at which the respondents had valued it

Complaints by the appellants started with the roof, which was the subject of a writ late in 1984, but over succeeding years up to the date of trial in 1987 other complaints were added, including movement of the structure and the need for underpinning - The sum claimed in the end was some £22,500, being the difference between the £87,500 paid and £65,000, alleged to be the value of the house in its defective condition

The judge at the trial found that the respondent firm had not been negligent under either of the main heads, subsidence or roof defects - As regards subsidence, the judge held that the appellants had in fact received the best possible advice, namely, to insure against the risk - They had done so and had received £15,700 towards the cost of underpinning, the insurers did not accept that all the underpinning proposed was necessary - The appellants thereby, recovered more than they would have been likely to obtain by other means -

As regards the roof, the judge held again that the respondents had not been negligent - The surveyor had inspected the inside of the roof and had seen no evidence of water penetration, such as would have put him on notice that a further investigation from the exterior was necessary, - It had been submitted to the court, although not pleaded, that the placing of roofing felt over the battens, which the surveyor had observed, should have alerted him to problems arising from lack of ventilation, possibly leading to deterioration of the slates, a substantial proportion of which were in fact subsequently found to be rotten, soft and powdery, - The judge, however, was not impressed by witnesses who agreed with this point when examined in court but had not referred to it in their reports.

It was argued for the appellants on appeal that the insurance was irrelevant; that the respondents had not made it clear that the cracks might require underpinning; and that the survey report was altogether too sanguine in its tone, encouraging the appellants to buy. a house which turned out to have serious defects - The Court of Appeal, however, agreed with the trial judge in rejecting the charges of negligence in regard to both roof and main structure - As regards subsidence, the surveyor had sounded a note of warning at every, point - He had made it clear that the house had defects characteristic of its age and situation - His concern was shown by his advice in regard to insurance, which resulted in a substantial contribution being made by. the insurers - As regards the roof, the court agreed that, as the inspection of the inside showed no evidence of water penetration, the surveyor was not negligent in failing to carry, the investigation further - Russell L J criticised the absence from the pleadings of the particular allegation of negligence based on failure to follow tip the consequences of positioning the roof felt over the battens - This was a matter pressed both before the trial judge and the Court of Appeal but it found no place in the pleadings - It was not a satisfactory procedure and it could reflect adversely on the weight to be attached to expert and other testimony - Appeal dismissed

Ezekiel and Another V McDade And Others
Official Referee's Business

Negligence - Valuation for building society - Liability in tort - Whether valuer negligent - Damages for reduction in value - Whether additional damages for worry, distress and inconvenience

Held:

Judgment was given for the plaintiffs. The first defendant was negligent; a reasonabiy competent surveyor making a "head and shoulders" inspection of the roof space would have observed the unacceptable 40mm gap and would not have reported, as the first defendant did, that there was no evidence of any serious defect in the examined parts of the property. Damages totalling £43,364.65 were awarded which included: damages based on the difference in value of the property in 1986 if it had been in the condition represented by the first defendant's report and its actual value were awarded in the sum of £36,000; damages for worry, distress and physical inconvenience of £6,000; and agreed damages of £1,364.65 for legal expenses and other costs.

Farley v. Skinner
[2000] EGCS 52

Full survey - request to advise on the effect of noise from nearby airport - failing to warn purchaser of noise of aircraft - quantum - damages for distress and inconvenience recoverable?

Claimant purchaser engaged the services of the Defendant surveyor to give a full report on a large residential property in East Sussex. The Claimant specifically requested that the Defendant was to advise on whether the property would be affected by aircraft noise from the nearby Gatwick Airport. The Defendant failed to discover that the property was a short distance from a navigation beacon used by air traffic control at Gatwick causing there to be excessive aircraft noise at the property. At first instance, the Defendant was found to have been negligent, a finding that was not challenged on appeal. The trial judge awarded the Claimant £10,750 as damages for impairment of the Claimant's use and enjoyment of the property.

Held:

(by a 2 to 1 majority) on appeal, that the law relating to the recovery of damages for stress and inconvenience in contractual claims had been succinctly stated in Watts v. Morrow. Such damages are irrecoverable in breach of contract cases unless the very essence of the contract was to provide pleasure, relaxation or peace of mind. The court ruled that a contract to survey the condition of a house did not fall within this exceptional category and thus the Claimant was ordered to pay back his award of damages.

First National Commercial Bank Plc v. Andrew S. Taylor Commercial Ltd [1997] PNLR 37

Valuation - development property - contributory negligence - lender's failure to make adequate enquiries of borrower - failure to mitigate - duty to monitor the loan account - duty to obtain reasonable price on resale.

Claimant advanced a loan to H. Ltd. in the sum of £525,000, following a valuation by the Defendant of £410,000, which was to be secured on a development property in London by way of a legal charge and personal guarantees by AB of the company and his brother CB. The loan was made subject to the Claimant first carrying out various status enquiries such as company searches and obtaining references from H Ltd.'s accountants, solicitors and bank. The Claimant failed to follow up on the details of charges over properties, which were revealed by the search. No steps were taken to clarify AB's relationship with the company when it was revealed that he was neither an officer nor a shareholder. No checks were made to verify AB's statement of assets and liabilities and no checks were made whatsoever about CB. After H Ltd. continually defaulted on the repayments the Claimant made a formal demand for payment. The guarantors offered £15,000 in settlement of their liabilities under the guarantee, which was accepted when it was discovered that they had no assets. The property was subsequently re-sold for £210,000.

Held:

The valuation given by the Defendant's came within the acceptable margin of reasonable valuations and thus there was no finding of negligence. In any event, had there been a finding of negligence against the Defendants, the award of damages would have been subject to substantial reductions on the basis of contributory negligence and a failure to mitigate on the part of the Claimant. The failure to make numerous enquiries of H Ltd. and AB in the face of suspicious circumstances, which had been apparent prior to the advance, amounted to contributory negligence, which would have been assessed at 75 per cent. Further deductions would have been made to account for the Claimant's failure to monitor the account properly, for reselling the property at an undervalue and for failing to actively pursue the guarantors effectively.

Gibbs and Another v Arnold Son & Hockley
Queen's Bench Division
[1989] 45 EG 156

Negligence alleged - Valuation report for loan on mortgage - Loan of 100% of purchase price for purchase of end-ofterrace Victorian house at £19,500 - Defendant firm instructed to value by Halifax Building Society, the valuation fee being paid by borrowers - Claim for negligence made against defendant firm after a builder's and an engineer's report - Whether defendants in breach of duly of care to plaintiff purchasers, as laid down in Smith v Eric S Bush and, if so, what was the quantum of damages - Judge found on the evidence that liability was not proved against defendants, so that the question of damages did not arise - Defendants' inspection and report had been done with great care - Firm a credit to surveyors' profession - Plaintiffs "ill-advised young couple"

Plaintiffs were a young couple who were about to be married before the purchase and were married and moved into the house soon after - Despite the usual advice by the building society, the plaintiffs decided not to arrange for a full structural survey; they were content to rely on the defendants' valuation report - That report had drawn attention to an old movement in the structure and the presence of an old and enlarged tie-bar to restrain the gable wall - Further movement was considered to be unlikely - Mention was made of deteriorating joinery, worn paintwork and need to repair windows as well as lack of insulation in the roof - A gutter needed replacement and a cement fillet around the chimney was required - However, the defendants valued the house at the agreed price of £19,500 and recommended the maximum advance

In the course of carrying out certain works (aided by, an improvement grant) the plaintiffs' builder encountered problems with a chimney stack and an engineer was called in - He produced a formidable-looking list of defects and recommendations, which prompted a claim for negligence against the defendants - The issues before the judge were whether the defects described in the engineer's report existed at the date of the defendants' inspection and, if so, whether they should have been discovered by the defendants - The answers turned on the judge's appraisal of the expert evidence, which was conflicting but tended as the case went on to indicate that the seriousness of the matters recorded in the engineer's report had been exaggerated - In the end, after a detailed examination of the evidence on all the matters complained of, the judge came to the conclusion that the case against the defendants had not been substantiated; their liability had not been established - The question of quantum of damages did not arise, but the judge made a few comments on items in the statement of claim

In the course of his judgment the judge expressed the view that anyone buying a house of this age should, in the absence of the vendors' providing an appropriate certificate to the contrary from an appropriate specialist firm, expect there to be small amounts of timber decay and beetle attack here and there which might need relatively inexpensive attention in years to come - In the present case the possibility of modest problems with the timber had been adequately covered in the defendants' report

In conclusion the judge said that he had formed an extremely high opinion of the representatives of the defendants' firm, who had been named in the case as professional men of very considerable knowledge, ability thoroughness and integrity - No slur should be cast on the good name of the firm "by this unfortunate and ill-advised young couple bringing this action" - Action dismissed

Glasgow Corporation v. Muir
[1943] 2 All ER 44

Negligence - Public tea room and shop - Shop crowded with children. Tea urn carried through with leave of manageress - Urn upset and children scalded - Liability of owners of premises.

The manageress of a public tea room, to which. access was obtained by way of a small shop, allowed two members of a picnic party to carry through the shop an urn containing several gallons of boiling tea, about a dozen children being then at the shop counter buying sweets and ices. For a reason which was not explained, the hold of one of the bearers slipped so that tea was spilt and scalded several children, on whose behalf an. action was brought against the owners of the premises:-

Held,

(1) that the manageress was bound to take such reasonable care as would avoid risk of injury to such persons' as she could reasonably foresee might be injured by failure to exorcise such care and that, in the circumstances, the standard of care owed by the. bearers of the urn was at least as high as that owed by the manageress.

(2.) that the careful carriage of the urn past the children involved no such danger as could have been foreseen by an ordinary reasonable person as a natural and probable consequence, so as to lead him to clear the children out of the passage ;

(3) that in any event the action must fail since it had not been proved what was the cause of the accident.

Decision of the First Division of the Court of Session, 1942 S. C.126, reversed.

Hedley Byrne & Co. v. Heller & Partners
[1964] AC 465
[House Of Lords.]

Negligence - Duty of care to whom? - Careless misrepresentation - Bankers - References regarding company given by bankers to other bankers at customer's request - Communicated to customers inquiring about company's credit worthiness - Express disclaimer of responsibility - Whether a special relationship creating duty of care - Whether action against bankers maintainable.

Banking - Reference by bankers - Careless misrepresentation – References not justified - Plaintiffs' financial loss –Whether a special relationship creating a duty of care - Whether action against bankers maintainable.

Held,

that a negligent, though honest, misrepresentation, spoken or written, may give rise to an action for damages for financial loss caused thereby, apart from any contract or fiduciary relationship, since the law will imply a duty of care when a party seeking information from a party possessed of a special skill trusts him to exercise due care, and that party knew or ought to have known that reliance was being placed on his skill and judgment. However, since here there was an express disclaimer of responsibility, no such duty was, in any event implied.

Merivale Moore plc and another v Strutt & Parker
Court Of Appeal

Negligence - Valuation - 46-year leasehold interest - Rental values - Yield rate - Whether rental values negligent – Whether advice on yield rate negligent without qualification

Held:

The appeal and cross-appeal were dismissed. There was no evidence as to the proper "bracket" of rental values before the judge; it was not open to him to find negligence on the basis of the disparity between the rental figures in the advice and the figures he thought would have been correct or justified alone, at least without explaining, in substantially more detail, why that conclusion followed in this case. The finding of negligence in respect of the rental values was reversed. The judge was entitled to accept a yield rate of 8.5%. At the trial there was no evidence to counter the opinions of the experts that the 7.5% yield should not have been given without qualification. It was no defence to claim that an obligation to give an opinion did not include the need to reason out a valuation. One of the roles of the yield figure is to insert into the valuation the appropriate level of risk; the fact that a client knows that there is some risk does not mean that he does not need advice as to its extent and level. On the evidence available to the judge, it was virtually inescapable as plainly established that the omissions of the qualifications on yield was negligent. Per Sir Christopher Staughton: The right figure for rental value was exceeded by 16% and, in those circumstances, it was open to the judge to find that the advice on rental value was outside the range of any competent surveyor.

The Mortgage Corporation v. Halifax (SW) Ltd. and Anscombe & Ringland Ltd.
[1999] Lloyd's Rep. PN 159

Valuation - alleged over-valuation by two separate valuers - self-certified loan - burden of proving alternative transaction - quantum of damages - contributory negligence - apportionment of liability between two valuers.

Claimant advanced a loan of £3 million to T secured by way of a first legal charge on the property situated at The Bishop's Avenue in Hampstead, London, following valuations of £5.25 million and £5 million from the First and Second Defendants respectively. The loan was calculated as 60% of the lower of the two valuations received. Following T defaulting on the repayments, the Claimant repossessed and resold the property for £2.51 million, accepted as the market rate at that time. The First Defendant accepted its negligence but the Second Defendant maintained that its valuation had not been negligent. Both D's claimed that C would have lent a larger percentage of the property's value if lower valuations had been given and that in any event failure to check T's ability to service the loan amounted to contributory negligence.

Held:

On the evidence, £4.8 million represented the highest valuation that would still come within the range of valuations of a competent valuer. Accordingly, the Second Defendant had been negligent in arriving at its figure of £5 million. Liability was to be apportioned equally between the Defendants given the influence that each had on the Claimant's decision to advance the loan. The burden of proving that the Claimant would have lent the same amount even if non-negligent valuations had been given was on the Defendants. This argument was not accepted, the court finding that the Claimant would have been lent £95,000 less if the valuations had been accurate (£4.15 million). Thus the Claimant's loss was £95,000. The Claimant had contributed to its loss by failing to make proper enquiries as to T's ability to service the loan and his repayment record under prior mortgages. The court made a finding of 20 per cent contributory negligence.

Mortgage Corporation v Lambert & Co and another
Chancery Division
[1999] 42 EC 138

Negligence - Mortgage valuation - Limitation Act 1980 - Whether loss accrued more than six years before writ - Whether period of limitation extended by section 14A of 1980 Act - Whether plaintiff had material that suggested negligent valuation more than three years before writ

Held:

From the date of the first default, at least until 4 October 1990, the owners' covenant to pay instalments was, in all probability, valueless, or if it had any value at all it was certainly not shown to be sufficient to make up the shortfall between the amount lent and the true value of the property. It followed that in accordance with Nykredit Mortgage Bank plc v Edward Erdman Group Ltd (No 2) [1998] 1 EGLR 99, the plaintiffs loss first accrued more than six years prior to the date of the writ. For the purposes of applying section 14A of the Limitation Act 1980 to extend the period of limitation, and in particular section 14A(10), the plaintiff was in receipt of estimated values of the property prior to the three years before the issue of the writ. However, the reports contained estimates by unqualified persons obtained for purposes other than valuation, and they did not amount to material that ought of itself to have caused the plaintiff to question the defendants' original valuation. Accordingly, the plaintiffs claim was not statute-barred.

Murphy v. Brentwood District Council
[1991] 1 AC 398; [1990] 2 All ER 269
House of Lords

Negligence - Duty of care to whom? - Local authority - Duty to pass building plans - Local authority passing plans for construction of house in reliance on negligent advice of independent contractor - Defective design of foundation causing extensive damage to house - Occupier selling house in damaged condition - Whether occupier entitled to damages for diminution in value of house and other losses incurred

Held,

allowing the appeal, that, while the principle in Donoghue v. Stevenson [19321 A.C. 562 applied to impose a duty on the builder of a house to take reasonable care to avoid injury or damage, through defects in its construction, to the persons or property of those whom he ought to have in contemplation as likely to suffer such injury or damage, that principle as stated extended only to latent defects; that, where a defect was discovered before any injury to person or health or damage to property other than the defective house itself had been done, the expense incurred by a subsequent purchaser of the house in putting the defect right was pure economic loss; and that to hold that a local authority, in supervising compliance with the building regulations or byelaws, was under a common law duty to take reasonable care to avoid putting a purchaser of a house in a position in which he would be obliged to incur such economic loss was an extension of principle that should not, as a matter of policy, be affirmed.

Nyckeln Finance Co Ltd v Stumpbrook Continuation Ltd
Queen's Bench Division
[1994] 33 EG 93

Negligence - Valuation for mortgage purposes - Reliance - Whether defendant liable for collapse in property market - Whether plaintiff mitigated its losses - whether contributory negligence

Held:

The plaintiff did rely on the valuation, which played a real and substantial part in inducing the plaintiff to make the loan. On the evidence the highest non-negligent valuation at the material time would have been £23.5m; had the plaintiff received that valuation there would have been no loan transaction. The plaintiff was contributorily negligent to the extent of 20% of the damages suffered by making the loan of 90% of the purchase price by not properly satisfying itself as to the reliability of the valuation. Applying Banque Bruxelles Lambert SA v Eagle Star Insurance Co Ltd the defendant's valuation was the cause of the plaintiff's loss, and this loss did not include that attributable to the collapse in the property market. The amount of the loss attributable to the collapse of the property market was the difference between the highest non-negligent valuation of £23.5m and a proper sale price of £5.7m. Damages awarded included loss of interest paid by the plaintiff in borrowing to refinance the loan and out-of-pocket expenses. The plaintiff had failed to accept an offer to purchase at a price of £17m and its losses should be limited to loss sustained on the basis of a sale at that price in August 1990 even if there were no liability for the losses attributable to the collapse of the market.

Patel and another v. Hooper & Jackson (a firm) [1998] EGCS 160

Valuation - property purchased in reliance on report - property uninhabitable - inability to resell - quantum - measure of damages - entitlement to general damages for discomfort and inconvenience of rented accommodation.

Claimant's purchaser took £83,000 loan from a building society in order to purchase home for £95,000. Claimants had relied on Defendant's house buyer's report and valuation carried out on the building society's instruction. The report highlighted the need for extensive repairs and renovation but gave a valuation of £90,000. After completion, a full structural survey concluded that the property was uninhabitable due to structural defects and damp problems. The Claimants could not finance the necessary works nor could they sell the property in that condition. He therefore had to live in rented accommodation whilst continuing to make his mortgage repayments. At first instance, the valuer was found to have been negligent, a proper valuation would have been £65,000. The court also found that the Claimant had been unable to sell the property. An award of £139,911 was made incorporating incurred mortgage interest and endowment policy contributions and general damages to compensate the relative discomfort of having to live in rented accommodation

Held:

On appeal, that in a case brought by the private purchaser of a house, such a purchaser is entitled to the reasonable costs of extricating himself from the purchase including the costs of alternative accommodation. But, as the trial judge had found that if properly advised the Claimants would have purchased an alternative property, they would have incurred similar mortgage repayments and endowment contributions in any event.

Thus damages in respect of these amounts were to be disallowed. The trial judge had been justified in awarding small awards in respect of general damages in this case.

Perry V Sidney Phillips & Son (a firm)
[1982] 263 EG 888
Court Of Appeal

Claim by purchaser of house against surveyors for breach of contract and negligence - Substantial defects not disclosed in survey report - Appeal from decision of deputy High Court judge who found negligence and assessed damages on the basis of cost of repairs at date of trial - After trial, and pending the appeal by surveyors, the plaintiff purchaser sold the house in its unrepaired state - Cost of repairs basis for assessment of damages no longer appropriate - Issues before the Court of Appeal were the correct measure of damages and date of assessment and whether there was a right to damages for "vexation"

Held

that the correct measure was the difference between the price paid for the house and its value at the date of purchase, damages being assessed at the date of breach, not date of trial – "Some compensation for inflation" added in the shape of interest at appropriate rate from date of breach until judgment and afterwards at judgment rate - Damages for "vexation" properly awarded by judge - Objection based on Liesbosch case rejected - Compensation under head of "vexation to be a modest recognition of "anxiety, worry and distress" arising from the physical consequences of the breach - Damages to be assessed by an official referee - Surveyors' appeal allowed in part

Philips v. Ward
[1956] 1 WLR 471; [1956] 1 All ER 874

Damages – Negligence – Surveyor - Property purchased in reliance on negligent report by surveyor - Measure of damages –"Value" - Cost of repair not recoverable – Damages measurable by difference in value of property as described and as it ought to have been described - Date of assessment – Quantum t of damages not to be altered by fall in value of money between date when cause of action accrued and date of hearing of action.

Held,

(1) that the proper measure of damages in this case was the difference in money between the value of the property in the condition described and its value as it should have been described, namely, £4,000 ; the proper measure of damages was not the cost of repair, which included other factors, such as tax relief and savings on the annual cost of upkeep from the replacement of new for old.

Measure of damages in cases where damage was caused by a tortfeasor or by tenant's breach of repairs covenant, and in Baxter v. Gapp & Co. Ltd. [1939] 2 KB. 271; 55 T.L.R. 739; [1939] 2 All E R. 752 distinguished.

Morris v. Sands (unreported, July 15, 1955) considered.

(2) That the damages must be assessed at the prices prevailing when the damage occurred – 1952 - and that figure could not in law be affected by a fall thereafter in the value of money.

Decision of official referee affirmed.

Platform Home Loans Ltd v Oyston Shipways Ltd and others
House Of Lords
[1999] 13 EG 119
[1998] Ch. 466; [1998] 4 All ER 252

Negligence – Valuation for non-status mortgage - Contributory negligence - Law Reform (Contributory Negligence) Act 1945 – Meaning of damage – whether damage overall transaction loss or part of loss attributable to valuation

Held:

The appeal was allowed. On the application of section 1 (1) of the Law Reform (Contributory, Negligence) Act 1945, the contributory negligence percentage of 20% should be deducted from the plaintiff's basic loss of £611,748, to arrive at £489,398. As this sum is less than the overvaluation of £500,000, the damages did not have to be further reduced under the principle in South Australia Asset Corporation v York Montague Ltd. [1996] 2 EGLR 93 – the SAAMCO principle.

Per Lord Cooke of Thorndon dissenting: Because of the SAAMC0 principle, the valuer is not liable to the lender for the full loss of £611,748, but only for the overvaluation of £500,000; this latter sum was the "damage" for the purposes of section 1(1) of the 1945 Act and from which the contributory negligence percentage of 20% should be deducted.

Roberts And Another v J Hampson & Co
Queen's Bench Division
[1988] 37 EG 110

Negligence - Mortgage valuation - Three schemes offered by building society, namely,

(1) where the valuer's client was the building society, although a copy of the valuation was provided to the purchaser,

(2) where the purchaser was the valuer's client, a copy of the valuation going to the society,

(3) a full structural survey –

In the present case the plaintiff purchasers, a young couple with very modest resources, chose the first scheme (previous negotiations for another property, adopting the second scheme, having failed) - Plaintiffs were seeking a 95% mortgage - They purchased the subject property relying on a valuation carried out for the building society under the first scheme by a valuer employed by the defendant firm, a copy of the valuation having been received by the plaintiffs - Plaintiffs paid the asking price, £19,000 for the property - The valuer's report, which certified to the building society that the property was suitable for the maximum advance, drew attention to a limited amount of dampness in the external walls and a certain amount of dry rot in the skirtings of a bedroom; it recommended that the plaintiffs should undertake to engage a reputable timber specialist to eradicate the rot - Unfortunately, a few months after the plaintiffs moved into the house indications appeared of dry. rot which proved to be a serious infestation and the present action was brought

There was no doubt on the authorities that the defendants owed the plaintiffs a duty of care - Although the first scheme contemplated a limited appraisal only, it was to be an appraisal by a skilled professional man, taking reasonable care in making the valuation - It was accepted that this limited appraisal did not require the lifting of carpets or moving of furniture as a matter of course, but if there is a specific ground for suspicion and the trail of suspicion leads behind furniture or under carpets the valuer must take reasonable steps to follow the trail until he has all the information which it is reasonable for him to have before making his valuation - In some cases a half hour's inspection may suffice, but in others it may be necessary, to spend two or three or more times as long, regardless of the fee - In the present case the question of duty, was not complicated by the existence of a disclaimer

The evidence disclosed a breach of the duty so defined - The defendants' valuer mentioned "a certain amount of rot" in the skirtings, but in fact the skirting board was badly infested at the time and the valuer failed to check whether it was, as he wrongly believed, localised - He did not test the wood blocks of the floor and appeared to have ceased to follow the "trail of suspicion" in the skirtings because of the intervention of a piece of furniture (which, from the layout of the room, did not appear to have been substantial) - He did not recommend the retention of any part of the loan, although if he had done so the plaintiffs would not have gone ahead

As to the quantum of damages, the value of the house with its defects at the time of the survey was £15,000 (it was eventually sold for £14,000) - The difference between the £15,000 and the price paid by the plaintiffs was, therefore, £4,000 - Plaintiffs were also entitled to £1,500 for disturbance and disruption, making a total of £5,500

Smith v Eric S Bush (A Firm)
Harris And Another v Wyre Forest District Council And Another
House of Lords
[19891 17 EG 68 and [1989] 18 EG 99
[1989] 2 All ER 514

Negligence - Valuations for mortgage - Valuer's responsibility in tort to purchaser who has relied on valuer's report to the knowledge of the latter - Relationship---akin to contract - Clear duty of valuer to exercise reasonable skill and care - Important decision of House of Lords affirming decision of Court of Appeal in favour of purchaser in Smith v Eric S Bush but declaring Court of Appeal in error in Harris v Wyre Forest District Council - Court of Appeal wrong in latter case in holding that a disclaimer prevented a duty of care from arising, thus defeating the Unfair Contract Terms Act 1977 - Comprehensive review of authorities - Yianni v Edwin Evans & Sons correctly decided - Discussion of general attitude today. of purchasers of ordinary houses, who do not in most cases arrange for a structural survey but justifiably assume that the house, having been valued by a qualified person, must be worth not less than the amount of the loan offered by a building society or other mortgagee - The valuer in these cases assumes responsibility. to both mortgagee and purchaser by agreeing to carry out a valuation for mortgage purposes, knowing that the valuation fee has been paid by the purchaser and that the valuation will probably be relied on by the latter to decide whether to make the purchase

The two appeals before the House, although differing in details, gave rise to the same three questions, namely:

(1) Whether the valuer instructed to value the house for mortgage, knowing that his valuation would probably be relied on by the prospective purchasers, owed them a duty. to exercise reasonable skill and care;

(2) Whether the disclaimer of liability was a notice within the Unfair Contract Terms Act 1977 and therefore subject to the requirement of reasonableness; and

(3) Whether, if so, the disclaimer in the present cases satisfied the requirement of reasonableness

Smith v Erie S Bush (a firm)

In this case the valuers, who had failed to check the chimney supports, had not disputed before the Court of Appeal the existence of a duty of care or the breach of it, but had relied on the disclaimer - The court held that the disclaimer was not in the circumstances fair and reasonable and so was made ineffective by, the 1977 Act - This view was challenged in the House, but the challenge was rejected - The House agreed with the Court of Appeal that in all the circumstances, including the fact that the subject property, was at the lower end of the market, the purchaser's reliance on the mortgage valuation without further survey, and the unequal incidence of loss and damage if the valuation went wrong, it would not be fair or reasonable to allow the disclaimer to take effect - The House accordingly dismissed the valuers' appeal - It may be noted that Lord Griffiths in his speech, after mentioning that the burden of proof was on the valuer to establish fairness and reasonableness, set out a list (not purporting to bc exhaustive) of matters to be taken into account by a judge when determining this difficult question

Harris v Wyre Forest District Council

The facts here were broadly, similar, the fact that the valuer was an "in-house" valuer in the employment of the Wyre Forest District Council mortgagees being immaterial - His failure was in not detecting and advising on the extent of settlement and instability in the property - In this case, however, the Court of Appeal, reversing the decision of Schiemann J, who had decided in favour of the purchasers, had taken the critical step of holding that the effect of the disclaimer was to prevent the duty of care from arising at all and ipso facto to remove anything on which the 1977 Act could bite - In the view of the House this was a fundamental error, a kind of tortfeasors' charter, which would "emasculate" the 1977 Act, removing the public's defence against the imposition of exclusion clauses - The answer to the Court of Appeal's construction was given by. Lord Griffiths in his "but for" test, earlier expounded by, Slade L J in Phillips Products Ltd v Hyland; in considering the primary. question as to whether there is a duty. of care and a breach of it, the disclaimer provision must be left out of account - The result of this analysis in the Wyre Forest case was that, the duty. of care and its breach having been established, the remaining question was whether it was reasonable under the 1977 Act to allow the local authority and its valuer to rely on the disclaimer - The House held that it was not and allowed the purchasers' appeal

The speeches deserve study, in detail by valuers, both for their interpretations of the law and their descriptions of practice - It should be noted that the House reserved their position in regard to valuation for mortgage of quite different types of property', such as industrial property', large blocks of flats or very expensive houses, to which different considerations might apply - A comment by Lord Templeman containing a hint of warning should also be noted:---Any increase in fees, alleged to be justified by the decision of this House in these appeals, will no doubt be monitored by. the appropriate authorities-

Sneesby v. Goldings (a firm) [1995] EGCS 201

Court Of Appeal

Negligence - Mortgage valuation report - Whether valuer negligent in failing to report missing chimney-breast - Whether damages based on diminution in value should assume all defects in building would be known to hypothetical purchaser

Held:

The appeal was dismissed and the cross-appeal was allowed. Whether the surveyor failed to exercise such care as was reasonable in all the circumstances in failing to observe and draw attention to the potential fault in the structure of the house resulting from the absence of a chimney-breast was a pure question of fact; it did not involve any point of general principle as to whether surveyors carrying out mortgage valuations are under a general duty to open cupboards. The judge was accordingly entitled to prefer the evidence of the plaintiffs to that of the defendants. Had the surveyor discovered the defective chimney-breast, that would have alerted him to the other defects in the building. A comprehensive survey would therefore have been likely and a hypothetical purchaser would have offered a price for the building in its actual condition having been informed of all the defects likely to affect its valuation. Damages had therefore been properly assessed as the difference between the value of the building in its assumed condition and the value in the bad condition with all defects which would have been in a comprehensive report. However, the judge had wrongly included £2,434 for repairs in addition to the diminution in value; this sum was deducted from the damages, although the diminution was increased by £1,000. Leave to appeal a decision of an official referee should be made in accordance with RSC Ord 58 r 4(b) and Ord 59 r 14(4) in the first instance to the court below.

South Australian Asset Management Corp v. York Montague Ltd [1997] AC 191

United Bank of Kuwait v. Prudential Property Services Ltd [1994] 30 EG 103

Nykredit Mortgage Bank plc v. Edward Erdman Group Ltd [1995] 1 EGLR 129

on appeal from Banque Bruxelles Lambert SA v. Eagle Star Insurance Ltd [1994] 2 EGLR 108

House of Lords

Damages - Measure of damages - Surveyor's report - Plaintiffs lending money on basis of surveyor's negligent valuation – Subsequent general fall in property market and default by borrowers in repayment of loans - Plaintiffs obtaining possession and selling properties at price below valuation - Whether loss attributable to market fall recoverable

Held,

dismissing the appeal in the first case and allowing the appeals in the second and third cases, that the duty of the defendants in each case, which was the same in tort as in contract, had been to provide the plaintiffs with a correct valuation of the property, namely the figure that a reasonable valuer would have considered it most likely to fetch if sold on the open market; that where a person was under a duty to take reasonable care to provide information on which someone else would decide on a course of action he was, if negligent, responsible not for all the consequences of the course of action decided on but only for the foreseeable consequences of the information being wrong; that the measure of damages was the loss attributable to the inaccuracy of the information suffered by the plaintiff through embarking on the course of action on the assumption that the information was correct; that in the first case the consequence of the valuation being wrong had been that the plaintiffs had had £10m. less security than they had thought; that if they had had that margin they would have suffered no loss and the whole loss had therefore been within the scope of the defendants' duty; that in the second case the consequences of the valuation being wrong had been that the plaintiffs had had £700,000 or £650,000 less security than they had thought; that the defendants had not been asked to advise on the risk of the borrower defaulting and any increased risk of default consequent on their overvaluation did not affect the scope of their duty to the plaintiffs; and that the damages should be reduced to the difference between their valuation and the correct valuation; and that in the third case the damages should similarly be reduced to the difference between the defendants' valuation and the true value of the property at the date of valuation

Wagon Mound No. 1
[1961] 1 All ER 404

On Appeal From The Supreme Court Of New South Wales.

Damages - Negligence – Remoteness - Damage directly caused by negligent act - Dangerous thing - Furnace oil -Spilled on harbour Waters – Fire – Causation - Damage to wharf - Foreseeability of consequences the effective test of liability - One criterion for determining liability and compensation.

Australia - New South Wales – Negligence - Furnace oil spilled on harbour waters – Fire - Foreseeability of consequences the test of liability - Remission of claim founded on nuisance.

Ship's Names – Wagon Mound

Held,

on the footing that the damage was the direct result of the escape of the oil, that, applying the test of foreseeability, the problems appellants who, as found by the trial judge, could not reasonably be expected to have known that the oil would catch fire, were not liable for the damage.

In re Polemis and Furness Withy & Co. Ltd. not followed.

Watts v. Morrow [1991] 1 WLR 1421; [1991] 2 EGLR 152

Damages - Measure of damages – Surveyor's report – Surveyor's report negligently failing to disclose defects - Property purchased in reliance on negligent report - Purchaser repairing defects - Whether cost of repairs recoverable - Whether damages measurable by difference in value as described and true value of defective property -Measure of damages for distress and inconvenience

On appeal by the surveyor on quantum of damages: -

Held,

allowing the appeal,

(1) that, in the absence of any warranty that the condition of the property had been correctly described by the defendant, there was no basis for awarding the cost of repairs; that the proper measure of damages was the sum

needed to put the plaintiffs in as good a position as if the contract had been properly performed; and that, accordingly, the financial loss of the plaintiffs was limited to the difference between the value of the property as it was represented to be and its value in its true condition

(2) That in the case of the ordinary surveyor's contract general damages were recoverable only for distress and inconvenience caused by physical consequences of the breach of contract; that such damages should be a modest sum for the amount of physical discomfort endured; and that, on the facts, £750 was an appropriate sum for each plaintiff

Yuen Kun-Yeu v. Attorney-General of Hong Kong [1987] 2 All ER 705

Privy Council [Appeal From The Court Of Appeal Of Hong Kong]

Negligence - Duty of care to whom? - Commissioner of Deposit-taking Companies - Statutory protection in Hong Kong for depositors - Commissioner registering company - Company's affairs allegedly conducted fraudulently, speculatively and to detriment of depositors - Whether commissioner in discharge of statutory powers owing duty of care to potential depositors - Deposit-taking Companies Ordinance (Laws of Hong Kong, 1983 rev., c. 328), ss. 10(1)(2)(e), 14(1)(d)

Held,

dismissing the appeal,

(1) that a close and direct relationship between the parties had to be established before liability in tort for negligence could arise, and in determining that question all the circumstances had to be taken into consideration including

the reasonable contemplation of harm being caused to the plaintiff by the defendant's failure to take reasonable care, although foreseeability of injury by itself was insufficient to create a duty of care; that although, on the plaintiffs' allegations, information about the company giving rise B to the suspicion that its business was being conducted to the detriment of its depositors was available to the commissioner but not to members of the public, the commissioner had no power under the Deposit-taking Companies Ordinance to control the management of deposit-taking companies, and there was no special relationship between the commissioner and the company and its managers which placed the commissioner under a duty to take reasonable care to prevent the company and its managers causing financial loss to future depositors with the company; and that, therefore, in all the circumstances the relations between the commissioner and the plaintiffs were insufficiently close and direct for him to owe them, as members of an unascertained class of potential depositors with deposit-taking companies in Hong Kong, a duty in performing his functions under the Ordinance to take reasonable care to prevent them suffering financial loss by reason of the fraudulent and improvident conduct of the company's affairs.

(2) That the commissioner was under no liability to the plaintiffs for negligent misrepresentation since the Ordinancc merely placed him under a duty to supervise deposit-taking companies in the public interest and imposed no special responsibility on him towards individual potential depositors, and that by registering the company and allowing it to remain registered the commissioner was in fact making no representation as to the creditworthiness of the company on which the plaintiffs were entitled to rely; and that, accordingly, the plaintiffs' statement of claim had properly been struck out as disclosing no reasonable cause of action

Index

Table of Cases

Statutes

Appendix 1